Smith Ely Jelliffe

The Flora of Long Island

Smith Ely Jelliffe

The Flora of Long Island

ISBN/EAN: 9783337269920

Printed in Europe, USA, Canada, Australia, Japan

Cover: Foto ©berggeist007 / pixelio.de

More available books at **www.hansebooks.com**

The Flora of Long Island

BY

SMITH ELY JELLIFFE, M.D., Ph.D.

Price, $1.00.

LANCASTER, PA.
The New Era Printing Company.
1899.

THE FLORA OF LONG ISLAND.

INTRODUCTION.

In presenting this Flora of Long Island the writer feels that a few words of explanation will make it clear, why, in a region of which so many local floras have been written, it has been deemed a matter of sufficient interest to prepare a flora of the whole island.

The reason lies in the peculiar features of botanical interest that are caused by the position of the island in relation to the mainland and the history of the growth of plant life on the Atlantic slope, as shown by the combined evidence of geology and botany on Long Island. To obtain a just conception of what these features of interest are a brief glance at the geological formation of the island, both present and past, may be of service.

GEOLOGY.

It is well known that Long Island is a terminal moraine of the great transatlantic ice sheet that stretched across the continent from Nantucket, which may be considered to be the extreme southeastern limit of the glacial action to Ohio and still further west.* F. J. H. Merrell has given a very excellent account of the geology of Long Island and his account is here freely used.†

"Long Island is about 120 miles long and varies from a mile to seven or eight miles in breadth. As a whole it is comparatively low and flat, but throughout there is a range of hills extending from its extreme western end at Bay Ridge, which are a continuation from similar hills of Staten Island, northeasterly to Roslyn and thence continuing to Montauk Point in a series of elevations, the most important of which are known as West, Dix, Comac, Bald and Shinnecock Hills. The average height of this chain is about 250 feet; but at some points it is greater. Harbor Hill at Roslyn is 384 feet above the tide; Jane's Hill 383 feet, Reuland's Hill 340 and Wheatley's Hill is 369 feet above the sea.

"There is also along the north shore an elevation which usually follows the contour of the numerous bays and inlets, varying in height

---

* Upham, North American Ice Sheet, *Amer. Journ. Science*, 28, 1897, pp. 81–92; Terminal Moraines of the North American Ice Sheet, *Amer. Journ. Science*, III., p. 18.

† *Annals of the N. Y. Academy of Science*, 3, 1883-1885, pp. 341-364.

from 30 to 200 feet and almost continuous from Astoria to Orient Point. These two sets of hills are the results of glacial action, and the more southern chain marks the southern limit of the drift." Merrell shows that this drift is quite thin in many places, and thus lacks in part one of the characteristic features of a true terminal moraine.

"South of the backbone, as the central range of hills is called, the surface is nearly level, gently sloping southward in an unbroken gravelly plain ; while between this ridge and the north shore is a second plain, with an elevation of 50 to 100 feet. From many of the deep bays on the north shore valleys extend through the hills in a southerly direction. These depressions, thirty in number, between East New York and River Head, average about twenty-five feet * in depth and are usually occupied by small streams flowing southward. These valleys Merrell thinks are the beds of rivers formed by the melting of the ice sheet in the Champlain Period."

"There are no important lakes or rivers on Long Island, but there are numerous ponds of clear cold water, without visible inlet or outlet. The existence of these ponds depends on the fact that in the stratified sands of the island, which are underlain by clays, a uniform water level or plane exists,† which rises northward from low tide level on the south shore at the rate of $2\frac{1}{2}$ feet per mile. The largest of these ponds is lake Ronkonkoma, which is three miles in circumference and has a maximum depth of 83 feet. Thus we find a long, low island with a prominent ridge nearer the north than the south with long, low, sloping hills to the sea on the south."

"Stretching to the east of Long Island are a number of smaller islands, Shelter, Robbins, Plum, Gull and Gardiner ; these are to be considered as possessing the same physical or geological characteristics. A still more interesting fact is that Block Island and Martha's Vineyard and Nantucket are thought to be of much the same character and that we here see the remnants now overlain with moraine drift of what was once a continuous tract of land."

"The inroads of the sea combined with the gradual depression of the coast has resulted in a submergence of portions of the fringe leaving the higher parts in the form of islands."‡

* E. Lewis, Jr., *Am. Jour. Science*, Series III., Vol. 13.

† Dana, Manual of Geology, p. 664.

‡ A. H. Hollick. Observations on the Geology and Botany of Martha's Vineyard, *Trans. N. Y. Acad. Sci.*, 13, 1893, p. 8.

N. S. Shaler, Report on the Geology of Martha's Vineyard, Seventh Annual Report U. S. Geol. Surv., 1885-86.

F. J. H. Merrell, Notes on Geology of Block Island, *Trans. N. Y. Acad. Sci.*, 15, p. 16.

Thus, rising above the surface of the ocean there is an interrupted chain of morainal deposits stretching from Cape Cod in Massachusetts, through Nantucket, Martha's Vineyard, Block Island, Long Island, Staten Island to New Jersey, these deposits making up the *greater part* of these islands, but, what is of more interest, *not all* of the islands, of which mention will be made later.

Soil.

Restricting the discussion to Long Island, the soil up to the backbone consists of glacial drift, varying in depth. This drift has been described by earlier writers to consist of two kinds: * 1. Till or drift proper, a heterogeneous mixture of gravel, sand and clay with boulders; 2. the gravel drift, a deposit of coarse yellow gravel and sand, brought to its present place by glacial and alluvial action, but existing near by in a stratified condition before the arrival of the glacier. This yellow gravel drift, which in a comparatively unaltered condition, forms the soil of the pine barrens of southern and eastern Long Island, and is exposed in section, now in many places on the island, notably in the sand pits outside of Prospect Park, in the cut of the Coney Island Railroad and in the brick yard in Huntington. It is moreover equivalent to and indeed identical with the yellow drift or preglacial drift of New Jersey, a formation of very great extent in the state. Merrell further gives us some interesting figures with reference to the thickness of the till in various places. In the hills near Brooklyn it attains its maximum depth, probably between 150 and 200 feet. In Calvary Cemetery a boring showed the drift to be 139 feet deep, and Mount Prospect, in Prospect Park, is probably all till and it is 194 feet in height. All along the north shore the till is much thinner; in some places there is none and it averages in thickness from a few feet to about six. At the extreme eastern end of the island there are mounds of till from 80 to 200 feet high between Nepeague Bay and Montauk Point. From a geological point of view the boulders are of interest, representing as they do most of the geological ages, having been brought down from New England and Canada. Underlying the till on the whole northern shore is the layer of yellow gravel, which varies greatly in different parts of the island. Various details of digging are given by Merrell in his admirable paper so extensively quoted.

Thus briefly recapitulating what is known of the soil of Long Island, we have the following facts: A limited area of crystalline rock, consisting of lamellated gneiss and schists, similar to Manhattens

* Merrell, l. c.

rocks at Hell Gate and about Astoria. The greater portion of the north side covered with till of various thicknesses, in some places the yellow gravel, or preglacial drift coming to the surface. On the slope of the southern hills a large amount of disintegrate till brought down by the streams and rain, a small surface area made up of the preglacial drift, and finally the alluvial sand and the various sand marsh components of the south shore.

A certain amount of attention has been given to the soil because it is so closely correlated with the historical geology of the island which has been so ably investigated by Drs. N. L. Britton and A. Hollick.

It would needlessly extend this account to go at length into the many valuable discoveries of Drs. Britton and Hollick which have served to prove that underlying the whole of Long Island and extending up along the chain to Cape Cod there are Cretaceous or post-Cretaceous strata. Yet the facts are of such interest and have such an important bearing on some features of the flora of the island that a brief quotation of some of their discoveries is desirable.

"During Cretaceous and Tertiary times a series of fresh water or estuary and marine deposits (clays, sands, gravels and marls) was laid down along the eastern borders of the North American Continent. About the close of the Miocene, or the beginning of the Pliocene, an era of elevation began, which finally raised them hundreds, in places thousands of feet, above their present level, forming a wide coastal plain, which extended over the whole area where we now find them and for a considerable distance eastward, into what is now part of the bed of the Atlantic Ocean. On the land side this plain was bounded by the crystalline and triassic rocks of Connecticut, southern New York, Pennsylvania and southward as may be seen by an examination of any good geological map of the eastern United States. The evidence of its extension around Massachusetts and Rhode Island are now almost obliterated, but there seems to be every reason to believe that its land limits were approximately the coast line of the present day. In fact, a small isolated portion of the old coastal plain still exists apparently in the region of Marshfield, Mass., as indicated by E. Hitchcock in 1841,* and recently by N. S. Shaler.† Further north than Massachusetts, so far as I am aware, it is not even indicated, and, except for the presence of the well-recognized submerged plateau off our eastern shores, all further trace of the former coastal plain is lost.

* Final Report Geol. Mass., 11, 427.
† Tertiary and Cretaceous Deposits in Eastern Massachusetts, *Bull. Geol. Soc. Am.*, 1, 443.

Its eastern limits where it formerly met the waters of the Atlantic Ocean were probably where we now find the borders of this plateau to be, namely, at the 100-fathom contour."

"Shortly after the advent of the Ice Age the elevation had reached its maximum. The rivers had previously cut deep valleys through the easily eroded material forming the coastal plain in their course to the sea and when the continental glacier, pushing its way southward and eastward, finally flowed over the edges and escarpments of the hard crystalline rocks onto the soft and incoherent strata of the coastal plain, it scooped it out to a great depth in places and then, either carrying it forward in mass or else pushing and squeezing it ahead in a great contorted ridge capped by the boulder till, finally left it as part of the terminal moraine. Wherever these conditions have prevailed we find the phenomena to be the same and Long Island may be considered as one of the grandest object lessons in this connection."

Just when the period of elevation ended and that of depression began, in fact, whether it was previous to of subsequent to that of greatest ice accumulation, is yet a matter of controversy between authorities, but in either case, on the retreat of the glacier, we may picture to ourselves the terminal moraine forming an elevated ridge extending through Staten Island, Long Island and the islands to the eastward, forming a continuous, more or less elevated land connection to the north and east, with what remained of the coastal plain sloping away from it on one side and a trough filled with the water from the melting glacier on the other.

"The present rate of coast subsidence is about two feet per century; at this rate six thousand years ago practically the whole of the area included within the present twenty-fathom contour would have been above sea level—only the deepest parts of the trough of the sound being below it."

"This area, as may readily be seen, includes the whole of Staten Island, Long Island, Block Island, Martha's Vineyard and Nantucket, besides a respectable portion of the submerged coasts eastward and southward. It is also probable that at least a part of this area to the eastward, which at the present time is lower than the twenty-fathom contour, has become disproportionately so in modern times by tidal scouring, and that it was actually and relatively higher formerly than now.

"Under these circumstances we should, therefore, have had, during a considerable period of time, a continuous strip of land, except for the river outlets, all the way from New Jersey to Massachusetts, separated by a body of water scooped out by the glacier, which, in its present depressed and widened condition, we now call Long Island Sound, but which was then a fresh-water lake or broad river."

"Bearing these conditions in mind we next have to consider the still further subsidence of the Champlain Period, the re-elevation of the Terrace Period and the depression which is again going on at the present day. It is evident that at some time during these oscillations of level the sea, having eaten away the coastal plain, finally reached the barrier of the terminal moraine, where it still remains as the connecting link between Long Island and Massachusetts. The moraine gave way in places, channels were formed and detached portions remained to form the islands which we recognize to-day as Block Island, Martha's Vineyard, Nantucket and the host of other lesser islands which stream out from the end of Long Island towards Cape Cod and the Rhode Island shore, while the eroded portions are represented by the great submerged ridges which are known as the Nantucket and other shoals."

"The vast time ratios formerly considered necessary by geologists are gradually but surely giving way to more moderate estimates and it is of interest to note that from six thousand to ten thousand years is the latest accepted calculation of the time which probably elapsed since the final recession of the glacier, by one of our most acute and conservative authorities*—a period which is about coincident with the probable time when the area bounded by the twenty-fathom contour was above the sea level. It is needless to point out that it also implies no subsequent submergence of the remaining portions of this land since the flora was established. In other words, Long Island, Block Island, Martha's Vineyard, Nantucket, etc., as we now know them, have not been submerged since the final retreat of the glacier and their separation into islands by the submergence of the intervening land is a comparatively modern phenomenon, due to the depression and erosion which are actually at work, and which have produced such conspicuous results during the historic period."

I have quoted this rather full argument because of its inherent importance and because I have felt that I could not better it in any way. We gain from it a twofold conception of the causes which have in part influenced the Long Island Flora and the key to the interpretation of some notable facts.†

*Warren Upham, Estimates of Geologic Time, *Am. Journ Science*, 14, 1893, p. 209.

† A. Hollick: Preliminary Contribution to our Knowledge of the Cretaceous Formation of Long Island and Eastward. Here is to be found a complete historical discussion of the whole matter.

Firstly, we have a geologic flora; and secondly, we have a residual flora which, during the six thousand to ten thousand years spoken of, was gradually forced northward by the sea and the subsidence and gradually encroached upon from the north by the third flora, consisting of plants which had come from the north before the ice sheet.

I. Geological Flora; Old Cretaceous Flora.

It is beyond the purpose of the present writer to discuss in full the occurrence of Cretaceous plants on Long Island. That they are found there has been definitely proven by Dr. A. Hollick.*

A brief glance at the maps of the coast from Cape Cod southward reveals, if one looks at the contours that have been made by soundings, that for a space of seventy-five miles from the coast there is a gradual drop about on a line with the general slope of the hills above the water. At the hundred-fathom limit, however, there is a sudden and immense drop of the water; it is seen that here is to be found the old Tertiary continental coast line that existed before the advance of the glacier from the north. The gradual submergence of the whole eastern portion of the continent has proceeded until now the only portion exposed of the continent of this part of North America that existed in the Tertiary times is found along the narrow belt of land lying at the base of the glacial drift, and, as before quoted, certain out-crops in New Jersey, certain portions in Long Island and underlying deposits of glacial drift of varying thickness in the islands stretching toward Cape Cod. This has been so well shown by Dr. A. Hollick that I cannot refrain from repeating his exact words.—" From a study of the existing geological and floral additions, as I have elsewhere attempted to demonstrate, the indications are that at the close of the Ice Age there was a continuous strip of land, except for certain river outlets, extending from what is now New Jersey to the southeastern New England coast, with a large body of fresh water occupying the deepest parts of what is now the basin of Long Island Sound. This strip consisted of an elevated portion along the northern border, formed by the terminal moraine left behind on the final retreat of the ice, and a plain region to the south, of varying width, representing what remained of the old Tertiary coastal plain, which formerly extended out to what is now the one hundred fathom contour."

I here append a list of the Cretaceous fossils found on Long Island as given by A. Hollick.†

---

* Plant Distribution as a factor in the Interpretation of Geological Phenomena, Trans. N. Y. Acad. Sciences, Vol. XII.

† Some Further Notes on the Geology of the North Shore of Long Island with their Distribution.

LIST OF CRETACEOUS PLANTS FOUND ON LONG ISLAND.

Serenopsis Kempii Hollick;
Salix proteaefolia v. flexuosa Lesq.
Salix purpuroides Hollick.
Juglans crassipes Heer.
Juglans arctica Heer.?
Ficus protogaea Heer?
Ficus Willisiana Hollick.
Protaeoides daphnogenoides Heer.
Laurus Plutonia Heer.
Laurus Omalii Sap. et Mar.
Laurus Newberryana Hollick.
Sassafras progenitor Newb. mss.?
Sassafras acutilobum Lesq.
Cinnamomum Sezannense Wat.
Diospyros rotundifolia Lesq.
Diospyros primaeva Heer.
Myrsine elongata Newb. mss.
Andromeda Parlatorii Heer.
Viburnum integrifolium Newb. mss.
Aralia transversinervia Sap. et. Mar.
Aralia patens Newb. mss.?
Aralia Nassauensis Hollick.
Myrtophyllum (Eucalyptus?) Geinitzi Heer.
Eucalyptus? nervosa Newb. mss.?
Dalbergia Rinkiana Heer.
Hymenaea Dakotana Lesq.?
Leguminosites constrictus Lesq.?
Leguminosites convolutus Lesq.?
Colutea primordiales Heer.
Sapindus Morrisoni Lesq.
Cissites formosus Heer.?
Paliurus integrifolius Hollick.
Zizyphus elegans Hollick.
Zizyphus Lewisiana Hollick.
Rhamnus? acuta Heer.
Celastrophyllum Benedeni Sap. et Mar.
Celastrophyllum decurrens Lesq.?
Grewiopsis viburnifolia Ward.
Menispermites Brysoniana Hollick.
Magnolia speciosa Heer.
Magnolia Capellini Heer.
Magnolia Isbergiana Heer.
Magnolia longipes Newb. mss.
Magnolia glaucoides Newb. mss.
Magnolia Van Ingeni Hollick.
Liriodendron primaevum Newb.
Liriodendron simplex Newb.
Liriodendron oblongifolium Newb. mss.
Tricalycites papyraceus Newb. mss.
Podozamites?
Poacites?
Cyperites?
Typha?

Making a list of 50 species distributed in 20 genera.

Another feature with reference to the history of the fossil plants of the island and which has not yet received its full share of attention is the occurrence of Tertiary deposits of diatoms. Merrell, in his Geology of Long Island, etc., was the first to mention the deposit found at Glen Cove, and more recently Dr. A. M. Edwards has obtained some rich findings from some deep cuts on the south side of the island.

As the evidence to be derived from diatoms is most conflicting, I refer those interested to the deposits found at Arverne by Dr. Edwards. (See Diatoms in list.)*

II. PINE BARREN FLORA.

In New Jersey there has been recognized for a number of years, a peculiar flora, known as the Pine Barren Flora; this was found to grow in a more or less restricted area, extending from New York to Cape May and the mouth of the Delaware River, occupying a narrow

---

*Also in note of H. Ries on Tertiary Clays at Glen Cove, Transactions Academy of Sciences, Vol. XIII., p. 167.

strip along the Atlantic coast. The geological strata, as pointed out by Dr. Britton,* upon which this flora grows are: Tertiary, lying to the south and southeast of a line drawn from a point on the Atlantic coast a few miles south of Long Branch, to another near the head of Delaware Bay; and Cretaceous, north of this line and extending between it and the southern edge of the Triassic formation, which follows a line from the center of Staten Island to the vicinity of Trenton.

"The Tertiary pine barrens extend southward along the Atlantic coast to Florida; it is with the flora of the northern extension of these sandy stretches of Cretaceous age which we have to do at present."

Dr. Britton shows that this flora, the northern pine barren flora, is characteristic, that it is found upon the yellow gravel drift and that it is a flora which is peculiarly American in its character. He also shows that a certain number of its plants are found in the southern portions of Staten Island, that it extends to Long Island and, the discussion of the geological features would tend to show, extends to the north, in Martha's Vineyard and Nantucket, which has been proven to be the case by Dr. Hollick and others.

The most important of these pine barren plants are here given.

LIST OF SPECIES.

Andropogon glomeratus.
" macrourus.
" hypericoides.
Ascyrum Crux Andreae.
" stans.
Aster spectabilis.
" nemoralis.
" concolor.
Pierus Mariana.
Asclepias obtusifolia.
" Caroliniana.
Arenaria squarrosa.

Crataegus uniflora.
Chrysopsis Mariana.
" falcata.
Coreopsis rosea.
Cyperus cylindricus.
Chamaecyparis thyoides.

Drosera filiformis.
Meibomia laevigata.
" viridiflora.

Eupatorium rotundifolium.
" album.
" hyssopifolium.
" leucolepis.
Euphorbia Ipecacuanhae.
Eleocharis melanocarpa.

Gnaphalium purpureum.
Gaylussacia dumosa.
Glyceria obtusa.

Hudsonia ericoides.
Helonias bullata.
Helianthus angustifolius.

Ipomoea pandurata.

Juncus pelocarpus.
" scirpoides var. macrostemon

Kalmia angustifolia.

Magnolia glauca.

*On the Northward Extension of the New Jersey Pine Barren Flora on Long and Staten Island: Bull. Torr. Bot. Club, 7, 1880, p. 81.

Polygala lutea
Phlox subulata.
Panicum verrucosum.

Quercus nigra.
" prinoides.
" Phellos.
" heterophylla.
" Rudkinii.

Rubus cuneifolius.

Solidago puberula.
Spiranthes simplex.

Stipa avenacea.
Sporobolus serotinus.

Tephrosia Virginiana.

Utricularia subulata.

Xyris flexuosa.
" Caroliniana.

Lycopodium inundatum.

Catharinea crispa.

W. W. Bailey, Pine Barren Plants in Rhode Island, *Bull. Torr. Bot. Club*, 7, 1880, p. 98. Lists nineteen of these species which were found in a limited area about Worden's Pond in southern Rhode Island.

E. W. Hervey, Flora of New Bedford and the Shores of Buzzard's Bay, with a Procession of the Flowers. Notes twenty-five of these plants.

Going northward their numbers become fewer and fewer until finally, as Hollick* has pointed out, some of the number originally reported by Britton may be excluded as they are found as far north as Canada; these are *Lycopodium inundatum, Asclepias obtusifolia, Juncus pelocarpus, Kalmia angustifolia, Solidago puberula* and *Tephrosia Virginiana*.

III. THE NORTHERN FLORA.

In the Cretaceous times, as already outlined, there was evidently quite an extensive flora on the continent of North America; we know from the researches of Newberry and Lesquereux that in this age we find the first angiosperms, so that the Cretaceous saw the dawn of the flora as we know it at the present day. With the advent of the glacial epoch many of the northern plants perished and the more hardy northern members of the flora slowly migrated southward, some of them, no doubt, living close to the margins of the ice sheet. At the close of the glacial epoch, many of the plants again migrated northward slowly and some of the distinctly southern plants moved up to the area of the terminal moraine. Many adapted themselves to the new soil and moved northward, but the probabilities seem to point to the fact that the set of plants we have designated as pine barren plants were enabled to hold their own only upon the yellow gravel, a

*l.c., Plant distribution, etc.

soil which they grew accustomed to and were unable to live upon the true glacial till.

Before the advent of man, the northern part of Long Island undoubtedly had a distinct forest growth ; the evergreens, hickories and chestnuts were probably abundant, some magnificent specimens of pines are still found in isolated patches on the islands, the remnants of probably a fine growth of these plants. When the trees were cut down and the lands tilled the present flora, which is continuous with that of New York and Connecticut and Massachusetts, became manifest. There are no features common to this north side of the island flora which is not reproduced in the southern parts of Connecticut and New York, save in the absence of the rock-loving plants ; hence, many of the ferns are absent, few of the rock lichens are found and almost none of the boulder spring flowers.

At the present time when the woods are undisturbed there are a number of chestnuts, sassafras, wild cherry, swamp-maple, locusts, cedars, hickory in places, a few tulip trees, some oaks, and in the cleared grounds the (aspen) rapidly gains a foothold.

In the woods the herbs are distinctly northern in their type and they usually run over the ridge and have extended to the southern borders of the island.

Along the sandy shores of the north side, of the more western portion of the island, the vegetation is identical with that on the southern shores of Connecticut.

There is abundant opportunity to study the sand beach flora on Long Island. It is developed to a great extent and throughout the eastern reaches of the island the vast sand dunes and wastes support a characteristic number of plants.

SOME STATISTICAL NOTES ON THE PRESENT FLORA.

It may be considered that the present cryptogamic flora of the island is very imperfectly known. The workers in this field have not been many and the regions explored have been very restricted. The establishing of the Biological Laboratory of the Brooklyn Institute at Cold Spring has given a start to the investigation of the cryptogamic flora which will probably yield many additions to the list. Of the Myxomycetes, there are noted some twelve species. *Lycogala Epidendron* and *Stemonitis fusca* seem to be the most widely distributed plants of the group. About 391 Algae are here listed, of these about 50 were desmids, 126 diatoms, and the remaining 215 members of the various other algal groups. The comparative poverty of the marine algal forms, perhaps better studied than any other of our cryptogamic plants has been made a special topic for discussion by Dr. W. G. Farlow, in his Presidental address before the Society of Plant Morphologists, held in December, 1898, at Columbia University.

The bacterial flora is absolutely unknown. Beyond the well recognized forms characteristic of certain diseased conditions in man and other animals our knowledge is limited. Some few investigations of the writer on the bacteria of the air of Brooklyn and the bacteria of the Brooklyn and Flushing water supplies yielded a number of forms which are not here listed because of the general lack of criteria for specific determinations, which lack is now fortunately being remedied by such workers as Chester, Dyar, Fuller and others, who are investigating the bacteria flora of the water supplies. Twenty-seven species alone are listed, but it may be recorded that at least three times that number have been isolated but not positively identified. The Phycomycetes are listed with only six species, probably a very meagre representation of this extensive group. The like pertains to the fifty-four species grouped with the Ascomycetes, 34 Hyphomycetes, or Fungi Imperfecti. The Basidiomycetes, by reason of their conspicuousness, contribute the greatest number to the list, there being some 136 species. There are eleven representatives of the Gasteromycetes. General remarks on the fungus flora are out of place by reason of the desultory character of the observations and the few workers. The lichens of the eastern end of the island were more carefully studied by the writer. No other workers in the same field have contributed to the literature, but collections by earlier observers were placed at my disposal, and it is with special pleasure that I am able to put on record in these higher cryptogamic groups the names of some of the earlier enthusiastic botanists of the old school. It has always been a matter of personal regret that I never knew them. Fifty-four species are here listed. The Bryophytes have received more attention. A list of twenty-seven Hepaticae in this restricted area is of interest. Howe is the only investigator who has thus far made any contribution to the study of these plants in the particular field. In the mosses 109 species are listed most of which has been collected by Miss M. L. Saniel.

In conclusion of this special feature, it may be hoped that this list will serve as a beginning for a number of the workers who are just now entering upon the study of the cryptogamic forms of the island.

The Pteridophytes are represented by forty-one forms, and fourteen Gymnosperms are listed. Three hundred and twenty-two Monocoty-

ledons are represented; of these, the large families, Graminaceae and Cyperaceae, still imperfectly explored, contribute ninety-nine and ninety-five species respectively.

In résumé then, the following list records 2238 species as occurring on Long Island, of which, following the older lines, 896 are Cryptogams and 1342 Phanerogams. These in tabular form may be thus represented:

Thallophytes	Algae	391
	Fungi	274
	Lichens	54
Bryophytes	Hepaticae	27
	Musci	109
Pteridophytes		41
Gymnosperms		14
Monocotyledons		322
Dicotyledons		1006

In conclusion the writer wishes to express his sincerest thanks to the many workers who have contributed so largely in the preparation of this list. It is impossible to name them all but the writer wishes in particular to acknowledge his indebtedness to Dr. N. L. Britton for constant help and guidance.

Partial Bibliography.

Contributions to the Botany of the State of New York. Chas. H. Peck.

Bulletin New York State Museum of Natural History. Different numbers contain references to Long Island stations.

Catalogue of Plants, Indigenous and Cultivated, found in the vicinity of Erasmus Hall, by John B. Zabriskie. 48th Ann. Rept. Regents, 176-181. 1835.

A Catalogue of Plants growing within thirty miles of Yale College, by D. C. Eaton. Published by Berzelius Society. 1874.

Plantae Plandomensis, or a Catalogue of the Plants growing spontaneously in the neighborhood of Plandome, the county residence of Samuel L. Mitchell, by Caspar Wistar Eddy, New York, August 28, 1807.

The Medical Repository, Vol. XI., pp. 123-131. New York. 1807.

Catalogue of Phaenogamous and Vascular Cryptogamous Plants of Queen's County, L. I., showing distribution through various townships. Julius A. Bisky, Ph.G., Flushing, L. I. MSS. Herbarium in Flushing High School.

Plants of Prospect Park, by Smith Ely Jelliffe, M.D., Curator of Herbarium of the Brooklyn Institute. The Brooklyn Daily Eagle Almanac, pp. 75, 76. 1890.

The Plants of Prospect Park. Additions to the list which appeared in the Eagle Almanac of 1890, by Smith Ely Jelliffe, M.D. The Brooklyn Daily Eagle Almanac, p. 94. 1890.

Notes on the Flora of Long Island. Science, N. Y., Vol. XXII., July 7, 1893.

The Shrubs and Trees of Prospect Park, by Col. Nicholas Pike. The Brooklyn Daily Eagle Almanac, 1892. Pp. 42, 43, 44. A thoroughly unreliable list.

Notes on the Flora of Long Island, by Rev. Geo. D. Hulst, Himrod Street, Brooklyn. MSS.

Suffolk County, Long Island Plants. 1887–1892. Named according to Gray's Manual. Edition of 1889–1890, by Mrs. L. D. Pychowska, Hoboken, N. J. MSS.

Catalogue of the Phaenogamous and Acrogenous Plants of Suffolk County, Long Island, by E. S. Miller, Wading River; and H. W. Young, Aqueboque P. O., L. I. Price, ten cents. Wm. A. Overton, Jr., & Co., printers, Port Jefferson, L. I., 1874.

Flora of New York State, by John Torrey, M.D., Albany, 1843. Contains a number of references to Long Island Plants.

Bulletin of the Torrey Botanical Club. (Miscellaneous contributions.)

Mss. contributed by D. Johnson, of Cold Spring Harbor.

Mss. contributed by Rev. G. D. Hulst, Ph.D.

Mss. contributed by Dr. M. A. Howe.

Mss. contributed by Miss M. Saniel.

Herbaria consulted: Torrey Club, Columbia University, Flushing High School, Brooklyn Institute, Long Island Historical Society, Herb. G. D. Hulst.

THALLOPHYTA.

MYXOTHALLOPHYTÀ—Myxomycetes.

CERATIOMYXA Schreb.
  C. mucida Pers.
    Queens: Cold Spring Harbor, Johnson.

TUBULINA Pers.
  T. cylindrica (Bull.) DC.
    Kings: Flatbush, Zabriskie. *Queens:* Cold Spring Harbor, Johnson.

DICTYDIUM Schrad.
  D. cernuum (Pers.) Nees.
    Kings: Flatbush, Zabriskie. *Queens:* Cold Spring Harbor, Johnson.

ARCYRIA Hill.
  A. cinerea (Bull.) Schum.
    Kings: Flatbush, Zabriskie.

LYCOGALA Micheli.
  L. epidendron Buxb.
    Frequent throughout.

TRICHIA Hall.
  T. varia Pers.
    Kings: New Lots.

HEMIARCYRIA Rost.
  H. rubiformis (Pers.) Rost.
    Kings: Bergen Island, Zabriskie; New Lots. *Queens:* Cold Spring Harbor, Johnson.
  H. clavata (Pers.) Rost.
    Kings: Flatbush, Zabriskie.

RETICULARIA Bulliard.
  R. Lycoperdon Bulliard.
    Queens: Cold Spring Harbor, Johnson.

COMATRICHA Preciss.
  C. longa Peck.
    Kings: Flatbush, Zabriskie.

STEMONITIS Gled.
  S. fusca Roth.
    Frequent throughout.

FULIGO Hall.
  F. varians Sommerf.
    Kings: Flatbush, Zabriskie.   *Queens:* Cold Spring Harbor, Johnson.

EUTHALLOPHYTA.

EUPHYCEAE.

CONJUGATEAE.

DESMIDIACEAE.

PENIUM Bréb.
  P. closteroides Ralfs.
    Frequent.

CLOSTERIUM Nitsch.
  C. gracile Bréb.
  C. Lunula Ehrb.
  C. striolatum Ehrb.
  C. Dianae Ehrb.
  C. Ehrenbergii Menegh.
  C. moniliferum Ehrb.
  C. rostratum Ehrb.
    These are frequent in the ponds of the island.

SPIROTAENIA Bréb.
  S. condensata Bréb.
    Ridgewood Water Supply.

PLEUROTAENIUM (Naeg.) Lund.
  P. Trabecula (Ehrb.) Naeg.
    Frequent in ponds.

DOCIDIUM (Bréb.) Lund.
  D. crenulatum (Ehrb.) Rab.
    Frequent.

COSMARIUM Corda.
  C. moniliforme Ralfs.
  C. Incus (Bréb.) Hass.
  C. Dianae Ehrb.
  C. Botrytis Menegh.
  C. Beckii Wille.
    These are all frequent in ponds throughout the island.

ARTHRODESMUS Ehrb.
  A. octocornis Ehrb.
    Ridgewood Water Supply.

HOLACANTHUM Lund.
  H. cristatum (Bréb.) Lund.
    Frequent.

XANTHIDIUM Ehrb.
  X. antilopaeum (Bréb.) Kutz.
    Ridgewood Water Supply.

STAURASTRUM Meyen.
  S. dejectum Bréb.
  S. brevispinum Bréb.
  S. aristiferum Ralfs.
  S. cuspidatum Bréb.
  S. brachiatum Ralfs.
  S. polymorphum Bréb.
  S. crenulatum (Naeg.) Delp.
  S. punctulatum Bréb.
  S. pentacladum Wolle.
  S. Sebaldi Reinsch.
  S. hirsutum (Ehrb.) Bréb.
  S. Arctiscon Ehrb.

  All more or less frequent in ponds throughout the island.

EUASTRUM Ehrb.
  E. circulare (Hass.) Ralfs.

  Frequent in ponds.

MICRASTERIAS Ag.
  M. Torreyi Bail.
  M. radiosa (Ag.) Ralfs.
  M. denticulata (Bréb.) Ralfs.
  M. fimbriata Ralfs.
  M. Americana (Ehrb.) Kütz.

  Frequent in ponds throughout the island.

SPHAEROZOSMA (Corda) Arch.
  S. filiforme Rab.
    Ridgewood Water Supply.
  S. spinulosum.
    Ridgewood Water Supply.
  S. excavatum Ralfs.
    Ridgewood Water Supply.

APTOGONUM Ralfs.
  A. Baileyi Ralfs.
    Common.

DESMIDIUM Ag.
  D. aptogonium Bréb.
    Ridgewood Water Supply.

HYALOTHECA Ehrb.
  H. dissiliens (Smith) Bréb.
    Ridgewood Water Supply.

ZYGNEMACEAE.

ZYGNEMA (Ag.) De Bary.
  Z. cruciatum Ag.
    Queens: Glen Cove.

SPIROGYRA Link.
  S. varians (Hass.) Kütz.
    Ridgewood Water Supply.
  S. nitida (Dill.) Link.
    Kings: Prospect Park.
  S. longata (Vauch.) Kütz.
    Kings: Prospect Park.
  S. setiformis (Roth) Kütz.
    Kings: Prospect Park.
  S. tenuissima Kütz.
    Ridgewood Water Supply.

ZYGOGONIUM Kütz.
  Z. pectinatum Kütz.
    Queens: Cold Spring Harbor, Johnson.

CHLOROPHYCEAE.

VOLVOCACEAE.

CHLAMYDOMONAS Ehrb.
  C. sp. indet.
    Common.

SPHAERELLA Sommerf.
  S. pluvialis (Ehrb.) Sommerf.
    Common throughout.

GONIUM Müll.
  G. pectorale Müll.
    Frequent throughout the island.

PANDORINA Bory.
  P. morum (Müll.) Bory.
    Kings: Prospect Park, Ridgewood Water Supply. *Queens:* Cold Spring Harbor, Johnson.

VOLVOX L.
  V. globator L.
    Not infrequent, local.

TETRASPORACEAE.

TETRASPORA Link.
  T. lubrica (Roth) Ag.
  T. bullosa (Roth) Ag.
    Frequent throughout.

DICTYOSPHAERIUM Naeg.
   D. Ehrenbergianum Naeg.
      Frequent.

PLEUROCOCCACEAE.

PLEUROCOCCUS Menegh.
   P. vulgaris Menegh.
      Common throughout.

RAPHIDIUM Kütz.
   R. polymorphum Fres.
   R. convolutum (Corda) Rab.
      Both frequent.

SCENEDESMUS Meyen.
   S. caudatus Corda.
   S. dimorphus Kütz.
      Both frequent.

POLYEDRIUM Naeg.
   P. longispinum Ehrb.
      Frequent.

STICHOCOCCUS Näg.
   S. bacillaris Näg.
      Queens: Cold Spring Harbor, Johnson.

HYDRODICTYACEAE.

PEDIASTRUM Meyen.
   P. Boryanum (Turp.) Menegh.
      Ridgewood Water Supply.
   P. pertusum Kütz.
      Ridgewood Water Supply.   *Queens:* Glen Cove.
   P. Ehrenbergii (Corda) A. Br.
      Kings: Prospect Park, Ridgewood Water Supply.

COELASTRUM Naeg.
   C. microporum Naeg.
      Ridgewood Water Supply.

HYDRODICTYON Roth.
   H. utriculatum Roth.
      Not infrequent throughout.

ULVACEAE.

MONOSTROMA (Thuret) Wittrock.
   M. Grevillei (Thuret) Wittr.
      Kings: College Point, Pike.   *Suffolk:* Sag Harbor, Pike.

M. pulchrum Farlow.
 Frequent.
M. latissimum Thur.
 Queens: Cold Spring Harbor, Johnson.

ULVA L.
 U. enteromorpha Le Jolis.
 Common along the coasts.
 U. Lactuca (L.) Le Jolis.
 Common along the coasts.
 U. clathrata Ag.
 Common along the coasts.
 V. Hopkirkii (McCalla) Harv.
 Queens: Cold Spring Harbor, Johnson.
 V. aureola (Ag.) Kütz.
 Queens: Cold Spring Harbor, Johnson.

ULOTHRICHACEAE.

ULOTHRIX Kütz.
 U. flacca (Dillw.) Thur.
 Frequent along the coasts.
 U. isogona (Engl.) Thur.
 Frequent along the coasts.
CONFERVA (L.) Lagehn.
 C. vulgaris Rab.
 Kings: Prospect Park.

CHAETOPHORACEAE.

DRAPARNALDIA Ag.
 D. glomerata Ag.
 In pools, quite commonly distributed.
CHAETOPHORA Schrank.
 C. pisiformis (Roth) Ag.
 Kings: Canarsie, Greenwood, Brainerd.
 C. endiviaefolia Ag.
 Kings: Greenwood, Brainerd.
TRENTEPOHLIA Ag.
 T. virgatula (Harv.) Farlow.
 Kings: College Point, Pike. Common in Long Island Sound, Farlow.
BULBOCOLEON Prings.
 B. piliferum Prings.
 Suffolk: Greenport, Pike.

OEDOGONIACEAE.

BULBOCHAETE Ag.
  B. intermedia De By.
    Kings: Ridgewood.

CLADOPHORACEAE.

CHAETOMORPHA Kütz.
  C. aerea (Dillw.) Kütz.
    New York Bay, Hooper, Brainerd.
  C. Linum (Fl. Dan.) Kütz.
    Common throughout the Sound.
  C. melagonium (Web. & Mohr.) Kütz.
    Queens: Rockaway Inlet, Pike.   *Suffolk:* Greenport, Pike.
  C. Picquotiana (Mont.) Kütz.
    Kings: Fort Hamilton, Pike.

RHIZOCLONIUM Kütz.
  R. salinum (Schleich.) Kütz.
    Not infrequent throughout.
  R. tortuosum Kütz.
    Suffolk: Orient, Pike,

CLADOPHORA Kütz.
  C. arcta (Dillw.) Kütz.
    Common throughout.
  C. albida (Huds.) Kütz.
    Kings: New York Bay, Harvey.   *Queens:* Jamaica Bay, Pike.
  C. lanosa (Roth) Kütz.
    Not infrequent throughout L. I. Sound.
  C. refracta (Roth) Aresch.
    Kings: New York Bay, Harvey.   *Queens:* Jamaica Bay, Pike.
  C. rupestris (L.) Kütz.
    Kings: Canarsie, Bath, College Point, Pike.
  C. glaucescens (Griff.) Harv.
    Not infrequent throughout the Sound.
  C. laetevirens (Dillw.) Harv.
    Kings: New York Bay, Harvey; Gravesend Bay, Pike.
  C. Hutchinsiae (Dillw.) Kütz.
    Kings: College Point, Pike.
  C. flexuosa (Griff.) Harv.
    Kings: Flushing Bay, Pike.
  C. Rudolphiana (Ag.) Harv.
    Kings: Hellgate, Pike.
  C. gracilis (Griff.) Kütz.
    Not infrequent.

C. expansa Kütz.
  Suffolk: Greenport, Pike.
C. fracta (Fl. Dan.) Kütz.
  Kings: New York Bay, Harvey, Walters, Pike.
C. glomerata Kütz.
  Kings: Prospect Park.

BRYOPSIDACEAE.

BRYOPSIS Lam.
  B. plumosa (Huds.) Ag.
  Common throughout.

VAUCHERIACEAE.

VAUCHERIA DC.
  V. Thuretii Woron.
  Reported as common by Pike, but the writer has never seen a specimen.
  V. sessilis (Vauch.) DC.
  Cold Spring Harbor, Johnson.
  V. litorea (Hoff.) Bang.
  Reported as common along coast by Pike; specimens so labelled examined by me have proven to be erroneously determined.

PHAEOPHYCEAE.

ECTOCARPACEAE.

ECTOCARPUS Lyngb.
  E. reptans Cron.
  Suffolk: Greenport, Port Jefferson, Pike. *Queens:* Cold Spring Harbor, Johnson.
  E. tomentosus (Huds.) Lyngb.
  Kings: College Point, Pike. *Queens:* Cold Spring Harbor, Johnson.
  E. granulosus (Eng. Bot.) Ag.
  Suffolk: Sag Harbor, Pike.
  E. confervoides (Roth) Le Jolis.
  Common throughout.
  E. fasciculatus Harv.
  Not infrequent throughout.
  E. lutosus Harv.
  Suffolk: Greenport, Harvey. Pike's specimens reported from Fort Hamilton and Great South Bay are young *E. littoralis*.
  E. littoralis Lyngb.
  Common throughout.

E. brachiatus Harv.
Kings: College Point, Jamaica Bay, Pike.

E. Hooperi Harv.[1]
Suffolk: Greenport, Harvey.

E. Dietzeae Harv.[1]
Suffolk: Greenport, Harvey.

SPHACELARIACEAE.

SPHACELARIA Lyngb.
S. cirrhosa (Roth) Ag.
Suffolk: Greenport, Pike.

S. radicans (Dillw.) Harv.
Queens: Cold Spring Harbor, Johnson.

CLADOSTEPHUS Ag.
C. verticellatus Ag.
Suffolk: Orient, Farlow; Pike's plants so determined are not this.

ENCOELIACEAE.

PUNCTARIA Grev.
P. latifolia Grev.
Common throughout.

P. plantaginea (Roth) Grev.
Not infrequent.

PHYLLITIS Kütz.
P. fascia (Fl. Dan.) Kütz.
Common throughout coast.

STRIARIACEAE.

STRIARIA Grev.
S. attenuata Grev.
Not infrequent throughout.

DESMARESTIACEAE.

ARTHOCLADIA Duby.
A. villosa Duby.
Suffolk: Greenport, Pike.

DESMARESTIA Lamx.
D. aculeata (L.) Lamx.
Kings: Flushing Bay, Brainerd; Fort Hamilton, Pike. *Suffolk:* Orient, Greenport, Pike.

D. viridis (Fl. Dan.) Lamx.
Common throughout Sound.

---

[1] See Notes on Algae of Long Island, W. G. Farlow, Bulletin Torrey Bot. Club, 20, 1893, pp. 107-109.

DICTYOSIPHONACEAE.

DICTYOSIPHON Grev.
  D. foeniculaceus (Huds.) Grev.
    Not infrequent along the whole coast.
  D. hippuroides (Lyngb.) Aresch.
    Kings: College Point, Jamaica Bay, Pike.

MYRIOTRICHIACEAE.

MYRIOTRICHIA Harv.
  M. clavaeformis Harv.
    Suffolk: Greenport, Orient, Montauk, Pike.
  M. radicans (Dillw.) Harv.
    Kings: College Point, Pike.

ELACHISTACEAE.

ELACHISTA Duby.
  E. fucicola (Velley) Fries.
    Frequent, though local.

CHORDARIACEAE.

MYRIONEMA Grev.
  M. vulgare Thur.
    Not infrequent along the whole coast.

CASTAGNEA (Derb. and Sol.) Thur.
  C. virescens (Carm.) Thur.
    Queens: Whitestone.   *Suffolk:* Orient, Pike.
  C. Zostereae (Mohr) Thur.
    Queens: Whitestone.   *Suffolk:* Orient, Pike.

MYRIACTIS Kütz.
  M. pulvinata Kütz.
    Suffolk: Greenport, Southampton, Orient, Pike.

LEATHESIA S. F. Gray.
  L. difformis (L.) Aresch.
    Not infrequent throughout.

MESOGLOIA Ag.
  M. divaricata (Ag.) Kutz.
    Not infrequent in bays along the south shore.

CHORDARIA Ag.
  C. flagelliformis (Fl. Dan.) Ag.
    Suffolk: Pike.

STILOPHORACEAE.

STILOPHORA Ag.
  S. rhizoides Ag.
    Not infrequent in the Sound.

LAMINARIACEAE.

CHORDA Stack.
  C. Filum (L.) Stack.
    Common throughout.

ALARIA Grev.
  A. esculenta Grev.
    Suffolk: Greenport, Orient, Peconic Bay, Pike.

LAMINARIA Lam.
  L. saccharina (L.) Lam.
    Found all along the coasts.
  L. digitata (L.) Lam.
    Suffolk: Montauk, Farlow, Pike's specimens from Greenport and Orient were not this.

FUCACEAE.

FUCUS L.
  F. vesiculosus L.
    Common throughout the island.
  F. ceranoides L.
    Reported by Pike from Fort Hamilton and Orient, but specimens seen were identical with others from foreign localities and apparently had been substituted.

ASCOPHYLLUM Stack.
  A. nodosum (L.) Le Jolis.
    Common throughout the northern shore.

SARGASSUM Ag.
  S. vulgare Ag.
    Commonly found throughout the coast, washed ashore.
  S. bacciferum (Turn.) Ag.
    Washed ashore throughout.

RHODOPHYCEAE.

BANGIACEAE.

BANGIA Lyngb.
  B. fusco-purpurea (Dillw.) Lyngb.
    Common throughout on piling.
  B. ceramicola (Lyngb.) Ch.
    Queens: Cold Spring Harbor, Johnson.

PORPHYRA Ag.
  P. laciniata (Light.) Ag.
    Common along the whole coast.

HELMINTHOCLADIACEAE.

BATRACHOSPERMUM Roth.
  B. moniliforme Roth.
    Frequent in fresh water throughout the island.

NEMALION Duby.
  N. multifidum Ag.
    Not infrequent throughout.

CHAETANGIACEAE.

SCINAIA Bivona.
  S. furcellata Bivona.
    Kings: College Point, Pike. *Suffolk:* Greenport, Pike.

GELIDIACEAE.

GELIDIUM Lam.
  G. crinale (Turn.) J. Ag.
    Common throughout.

GIGARTINACEAE.

CHONDRUS Stack.
  C. crispus (L.) Stack.
    Common throughout.

GIGARTINA Lam.
  G. mamillosa (Good. & Woodw.) J. Ag.
    Suffolk: Greenport, Montauk, Pike.

PHYLLOPHORA Grev.
  P. Brodiaei (Turn.) J. Ag.
    Pike's specimens from Glen Cove and Orient seem to be *P. membranifolia.*
  P. membranifolia (Good. & Wood.) Ag.
    Not infrequent throughout.

GYMNOGONGRUS Mart.
  G. Torreyi (Ag.) J. Ag.
    New York Bay, Agardh; Fort Hamilton, J. Hooper.

AHNFELDTIA Fr.
  A. plicata (Huds.) Fr.
    Frequent throughout.

RHODOPHYLLIDACEAE.

CYSTOCLONIUM Kütz.
  C. purpurascens (Huds.) Kütz.
    Common throughout.

EUTHORA J. Ag.
   E. cristata J. Ag.
      Infrequent throughout the coast.
RHABDONIA Harv.
   R. tenera Ag.
      Common along the coasts.

SPHAEROCOCCACEAE.

GRACILARIA Grev.
   G. multipartita J. Ag.
      Common along the coasts.
HYPNEA Lam.
   H. musciformis (Wulf.) Lam.
      Suffolk: Orient, Miss Booth.
RHODYMENIA Grev.
   R. palmata (L.) Grev.
      Not infrequent along the coasts.
LOMENTARIA (Gaill.) Thuret.
   L. uncinata Menegh.
      Common along the coasts.
   L. rosea (Harv.) Thuret.
      Suffolk: Montauk, Orient, Port Jefferson, Pike.
CHAMPIA Ag.
   C. parvula (Ag.) Harv.
      Common along the coasts.

DELESSERIACEAE.

GRINNELLIA Harv.
   G. Americana (Ag.) Harv.
      Common along the coasts.
DELESSERIA Lam.
   D. sinuosa (Good. & Woodw.) Lam.
      Not infrequent along the coasts.
   D. alata (Huds.) Lam.
      Suffolk: Montauk, Pike.
   D. Leprieurii Mont.
      Not infrequent along the coasts.

RHODOMELACEAE.

CHONDRIA (C. Ag.) Harv.
   C. dasyphylla (Woodw.) C. Ag.
      Suffolk: Southampton, Pike.
   C. tenuissima (Good. & Woodw.) C. Ag.
      Common along the coasts.

C. littoralis (Harv.) C. Ag.
 Kings: Fort Hamilton, Pike.
C. atropurpurea (Harv.) C. Ag.
 Queens: Glen Cove, Pike.

POLYSIPHONIA Grev.
 P. urceolata (Dillw.) Grev.
  Common along the coasts.
 P. subtilissima Mont.
  Queens: Jamaica Bay, Pike.
 P. Olneyi Harv.
  Kings: Canarsie, Pike.
 P. Harveyi Bailey.
  Not infrequent along Long Island.
 P. elongata Grev.
  Queens: Glen Cove, Pike.
 P. fibrillosa (Dillw.) Grev.
  Kings: Jamaica, Fort Hamilton, Pike.
 P. violacea (Roth) Grev.
  Suffolk: Greenport, Montauk Point, Pike.
 P. variegata (Ag.) Zan.
  Common along the coasts.
 P. parasitica Grev.
  Kings: Fort Hamilton, Pike.
 P. atrorubescens (Dillw.) Grev.
  Suffolk: Orient, Miss Booth; Noank, Farlow.
 P. nigrescens (Dillw.) Grev.
  Common along the coasts.
 P. fastigiata (Roth.) Grev.
  Queens: Glen Cove, Pike. *Suffolk:* Montauk Point, Pike.

BOSTRYCHIA Mont.
 B. rivularis Harv.
  Kings: Hellgate, Harvey; College Point, Peck; Rockaway Beach. *Queens:* Cold Spring Harbor, Johnson.

RHODOMELA C. Ag.
 R. subfusca (Woodward) C. Ag.
  Not infrequent along the whole coast.

DASYA Ag.
 D. elegans (Mart.) Ag.
  Common along the whole coast.

CERAMIACEAE.

SPERMOTHAMNION Aresch.
 S. Turneri (Mart.) Aresch.
  Not infrequent in Sound and in harbors of the south shore.

GRIFFITHSIA Ag.
  G. Bornetiana Farlow.
    Common along the eastern end of the island, infrequent about Brooklyn.

HALURUS Kütz.
  H. equisetifolius Kütz.
    Reported by Hooper from Brooklyn, probably Bay Ridge. *Suffolk*: Greenport and Great South Bay, Pike.

CALLITHAMNION Lyngb.
  C. Rothii (Eng. Bot.) Lyngb.
    Kings: Fort Hamilton, Pike.
  C. cruciatum Ag.
    Not infrequent about Brooklyn, now most of the stations destroyed. *Queens*: Cold Spring Harbor, Johnson.
  C. floccosum Ag.
    Suffolk: Orient, Pike.
  C. Pylaisaei Mont.
    Suffolk: Orient, Miss Booth.
  C. Americanum Harv.
    Common throughout.
  C. Plumula (Ellis) Lyngb.
    Suffolk: Orient, Miss Booth.
  C. Borreri (Eng. Bot.) Harv.
    Common throughout.
  C. roseum (Roth) Harv.
    Not infrequent throughout.
  C. polyspermum Ag.
    Kings: Hellgate, Harvey, Bath. *Queens*: Hempstead Bay, Pike.
  C. tetragonum (With.) Ag.
    Rare throughout region.
  C. Baileyi Harv.
    Common throughout.
  C. byssoideum Arn.
    Not infrequent throughout.
  C. corymbosum (Eng. Bot.) Lyngb.
    Suffolk: Greenport, Harvey, Pike.
  C. Dietziae Hooper.
    See notes by Farlow: Bull. Torrey Botanical Club, **20**, 1893, pp. 107–109.
  C. seirospermum Griff.
    Not infrequent throughout the island.
  C. ? tenue (Ag.) Harv.
    Not infreqent throughout the island.

PLUMARIA Stack.
- **P. elegans** (Bonnem) Schmitz.
  Common throughout.

PTILOTA C. Ag.
- **P. plumosa** (L.) C. Ag.
  Kings: Bath. *Suffolk:* Greenport, Pike.

SPYRIDIA Harv.
- **S. filamentosa** Harv.
  Not infrequent along the coasts.

CERAMIUM Roth.
- **C. rubrum** Ag.
  Common along the coasts.
- **C. circinnatum** Kütz.
  Queens: Glen Cove, Young.
- **C. diaphanum** (Lightf.) Roth.
  New York Bay, Harvey, Young.
- **C. strictum** (Kütz.) Kaw.
  Common along the coasts.
- **C. fastigiatum** Harv.
  Not infrequent about the coast of the Sound.
- **C. Hooperi** Harv.
  New York, Young.
- **C. tenuissimum** (Lyngb.) Ag.
  Common throughout the Sound.
- **C. Capri-Cornu** (Reinsch) Farlow.
  Kings: Canarsie, Young.

RHIZOPHYLLIDACEAE.

POLYIDES Ag.
- **P. rotundus** (Gmel.) Grev.
  Suffolk: Sag Harbor, Montauk, Pike.

SQUAMARIACEAE.

PEYSONNELIA Dec.
- **P. Dubyi** Crouan.
  Suffolk: Greenport, Peconic Bay, Pike.

CORALLINACEAE.

MELOBESIA Lam.
- **M. farinosa** Lam.
  Queens: Glen Cove, Pike; *Suffolk:* Greenport, Pike.
- **M. pustulata** Lam.
  Kings: Flushing Bay, Rockaway, Pike. ?
- **M. macrocarpa** Rosanoff.
  Kings: Canarsie; *Suffolk:* Orient, Quogue, Pike.

CORALLINA L.
  C. officinalis L.
    Stated to be common by Pike, but is rare.

HILDENBRANDTIA Wardo.
  H. rosea Kütz.
    Frequent throughout.

PERIDINIALES.
GYMNODINIACEAE.

GYMNODINIUM Stein.
  G. fuscum (Ehrb.) Stein.
    Brooklyn Water Supply.

PERIDINIACEAE.

CERATIUM Schrank.
  C. tripos Nitzsch.
    Brooklyn Water Supply, Bates. *Queens:* Cold Spring Harbor, Johnson.
  C. longicorne Perty.
    Brooklyn Water Supply.

PERIDINIUM Ehrb.
  P. tabulatum Ehrb.
    Brooklyn Water Supply.

CYANOPHYCEAE.
CHROOCOCCACEAE.

CLATHROCYSTIS Henfry.
  C. aeruginosa Henfry.
    Queens: Glen Cove.
  C. roseo-persicina Cohn.
    Queens: Cold Spring Harbor, Johnson.

COELOSPHAERIUM Naeg.
  C. Kützingianum Naeg.
    Frequent.

MERISMOPEDIA Meyen.
  M. glauca Naeg.
    Not infrequent.

NOSTOCACEAE.

NOSTOC Vauch.
  N. sphaericum Vauch.
    Kings: Greenwood, Hooper.

OSCILLARIACEAE.

OSCILLARIA Kütz.
  O. Froelichii Kütz.
    Frequent.
  O. aeruginea-caerulea Kütz.
    Frequent.
  O. limosa Kütz.
    Queens: Jamaica Bay, Whitestone, Pike.

MICROCOLEUS Desmaz.
  M. chthonoplastes (Fl. Dan.) Thuret.
    Kings: Fort Hamilton, Pike. *Suffolk:* Greenport, Pike. *Queens:* Cold Spring Harbor, Johnson.

SPIRULINA Turpin.
  S. tenuissima Kütz.
    Kings: Fort Hamilton, Gravesend, Pike. *Queens:* Cold Spring Harbor, Johnson.
  S. Thuretii Con.
    Queens: Cold Spring Harbor, Johnson.

SPHAEROZYGA Ag.
  S. polysperma Kütz.
    Queens: Cold Spring Harbor, Johnson.

LYNGBYA Ag.
  L. aestuarii (Jürg.) Laelm.
    Kings: Bay Ridge, Pike. *Queens:* Cold Spring Harbor, Johnson.
  L. luteofusca (Ag.) J. Ag.
    Not infrequent throughout coast.
  L. majuscula (Dill.) Harv.
    Kings: Canarsie, College Point, Pike.
  L. nigrescens Harv.
    Not infrequent throughout coast.
  L. tenerrima Thuret.
    Kings: College Point, Pike.

NODULARIA Martens.
    Kings: Bay Ridge, Hellgate, Pike.

RIVULARIACEAE.

CALOTHRIX Ag.
  C. convervicola (Roth) Ag.
    Not infrequent.
  C. crustacea (Sch.) B. & Thuret.
    Kings: Bay Ridge, Pike.

C. parasitica Thuret.
 Queens: Jamaica Bay, Pike. *Suffolk:* Greenport, Pike.
C. pulvinata (Mert.) Ag.
 Suffolk: Greenport, Pike.
C. scopulorum (Web. & Mohr) Ag.
 Suffolk: Greenport, Pike. *Queens:* Cold Spring Harbor, Johnson.

RIVULARIA Roth.
 R. atra Roth.
 Kings: Hell Gate, Flushing, Pike. *Queens:* Cold Spring Harbor, Johnson.
 R. nitida Ag.
 Queens: Jamaica Bay, Pike.

ISACTIS Thuret.
 I. plana (Harv.) Thuret.
 Kings: Fort Hamilton, Pike. *Queens:* Jamaica Bay, Pike.

BACILLARIALES.

DIATOMACEAE.[1]

MELOSIRA Ag.
 M. Borrerii Grev.
 Kings: Bay Ridge.
 M. granulata (Ehrb.) Ralfs.
 Frequent throughout the island. Fossil at Cold Spring.
 M. nummuloides (Bory) Ag.
 Kings: Bay Ridge.
 M. sulcata (Ehrb.) Kütz.
 Queens: Arverne, Edwards.
 M. varians Ag.
 Kings: Prospect Park, Brooklyn Water Supply.

HYALODISCUS Ehr.
 H. Franklinii Ehr.
 Queens: Arverne, Edwards.
 H. stelliger Bailey.
 Queens: Arverne, Edwards.
 (**Pyxidicula compressa** Bail.)
 Arverne, Edwards.

CYCLOTELLA Kütz.
 C striata (Kütz.) Grun.
 Queens: Arverne, Edwards.

---

[1] In this citation of the species Schütts' classification in Engler and Prantl's system is followed for the sake of general conformity although we prefer Van Heurck's arrangement in A Treatise on the Diatomaceae.

COSCINODISCUS Ehrb.
  C. excentricus Ehrb.
    Queens: Arverne, Edwards.
  C. asteromphalus Ehrb.
    Queens: Arverne, Edwards.
  C. lineatus Ehrb.
    Queens: Arverne, Edwards.
  C. nodulosus Grün.
    Queens: Arverne, Edwards.

ACTINOPTYCHUS Ehrb.
  A. undulatus Ehrb.
    Queens: Arverne, Edwards.

AULACODISCUS Ehrb.
  A. Germanicus Ehrb.
    Queens: Arverne, Edwards.

ACTINOCYCLUS Ehrb.
  A. Ehrenbergii R.
    Queens: Arverne, Edwards.

EUPODISCUS Ehrb.
  E. radiatus Bail.
    Kings: Coney Island, Woodman.

AULISCUS Ehrb.
  A. caelatus Bailey.
    Queens: Arverne, Edwards.
  A. pruinosus Bailey.
    Queens: Arverne, Edwards.
  A. radiatus Bailey.
    Queens: Arverne, Edwards.

RHIZOSOLENIA Ehrb.
  R. gracilis Ehrb.
    Frequent.

TRICERATIUM Brightwell.
  T. alternans Bailey.
    Kings: Coney Island, Woodman.   *Queens:* Arverne, Edwards.
  T. favus Ehrb.
    Queens: Arverne, Edwards.
  T. punctatum Brightwell.
    Queens: Arverne, Edwards.
  T. maculatum Kütz.
    Queens: Arverne, Edwards.

BIDDULPHIA Gray.
  B aurita (Lyngb.) Breb.
    Queens: Arverne, Edwards.

B. laevis Ehrb.
   Kings: Bay Ridge.
B. pulchella Gray.
   Queens: Arverne, Edwards.   *Suffolk:* Shelter Island.
B. Rhombus (Ehrb.) W. Sm.
   Queens: Arverne, Edwards.
B. turgida W. Sm.
   Queens: Arverne, Edwards.
B. Smithii (Ralfs) V. H.
   Queens: Arverne, Edwards.

ISTHMIA Ag.
   I. enervis Ehrb.
   Queens: Arverne, Edwards.

EUNOTOGRAMMA Weisse.
   E. Amphioxys Ehrb.
   Queens: Arverne, Edwards.

TERPSINOË Ehrb.
   T. Americana Ralfs.
   Queens: Arverne, Edwards.

RHABDONEMA Ehrb.
   R. Adriaticum Kütz.
   Not infrequent throughout the coast.
   R. arcuatum (Ag.) Kütz.
   Queens: Arverne, Edwards.

STRIATELLA Ag.
   S. interrupta (Ehrb.) Heiberg.
   Brooklyn Water Supply.

TABELLARIA Ehrb.
   T. fenestrata Kütz.
   Frequent throughout the island.   Freshwater.
   T. flocculosa (Roth) Kütz.
   Frequent throughout the island.

GRAMMATOPHORA Ehrb.
   G. marina Kütz.
   Kings: Bay Ridge.   *Queens:* Arverne, Edwards.   *Suffolk:* Shelter Island.

MERIDION Ag.
   M. circulare Ag. Fr.
   Brooklyn Water Supply.

DIATOMA DC.
   D. hiemale (Lyngb.) Heiberg.
   Queens: Glencove, Oyster Bay, Fossil, Ries.

PLAGIOGRAMMA Grev.
  P. **Gregorianum** R.K.E.
   Queens: Arverne, Edwards.
FRAGILARIA Lyngb.
  F. **capucina** Desmaz.
   Common throughout the island.
  F. **Pacifica** Grün.
   Queens: Arverne, Edwards.
SYNEDRA Ehrb.
  S. **affinis** Kütz.
   Not infrequent along the coasts.
  S. **fulgens** (Kütz.) W. Sm.
   Suffolk: Shelter Island.
  S. **pulchella** Kütz.
   Common throughout the island.
  S. **longissima** W. Sm.
   Brooklyn Water Water Supply.
  S. **lanceolata** Kütz.
   Kings: Prospect Park.
  S. **Ulna** (Nitzsch) Ehrb.
   Frequent throughout the island.
ASTERIONELLA Hass.
  A. **formosa** (Hass.) Fr.
   Frequent throughout the island.
EUNOTIA Ehrb.
  E. **lunaris** (Ehrb.) Grün.
   Frequent throughout the island.
  E. **major** (W. Sm.) Rab.
   Kings: Coney Island, Woodman.
  E. **monodon** Ehrb.
   Queens: Arverne, Edwards.
  E. **tridentula** Ehrb.
   Brooklyn Water Supply.
ACHNANTHES Bory.
  A. **brevipes** Ag.
   Kings: Bay Ridge.
  A. **longipes** Ag.
   Kings: Bay Ridge.
  A. **subsessilis** Ehrb.
   Kings: Bay Ridge.   *Queens:* Arverne, Edwards.
COCCONEIS Ehrb.
  C. **Pediculus** Ehrb.
   Frequent throughout the island.

C. scutellum Ehrb.
Frequent throughout the island.

NAVICULA Bory.

N. Americana Ehrb.
Kings : Coney Island, Woodman.

N. ambigua Ehrb.
Kings : Prospect Park.

N. Apis Kütz.
Kings : Coney Island, Woodman.

N. clavata Grün.
Queens : Arverne, Edwards.

N. flamma A. Schm.
Kings : Prospect Park.

N. cuspidata Kütz.
Frequent throughout the island.

N. didyma Ehrb.
Queens : Arverne, Edwards.

N. dilatata Ehrb.
Brooklyn Water Supply.

N. elliptica Kütz.
Queens : Arverne, Edwards.

N. firma Kütz.
Kings : Prospect Park, Coney Island, Woodman.

N. laevissima Kütz.
Kings : Prospect Park.

N. gibba (Kütz.) Ehrb.
Brooklyn Water Supply.

N. Hennedyi W. Sm.
Queens : Arverne, Edwards.

N. Hitchcockii Ehrb.
Kings : Prospect Park.

N. Indica Grev.
Queens : Arverne, Edwards.

N. lacustris W. Sm.
Queens : Arverne, Edwards.

N. lata Bréb.
Kings : Coney Island, Woodman.   *Queens :* Arverne, Edwards.

N. Lyra Ehrb.
Not infrequent around the coast.

N. major Kütz.
Kings : Prospect Park.

N. peregrina (Ehrb.) Kütz.
Queens : Arverne, Edwards.

N. permagna Bailey.
Queens : Arverne, Edwards.

N. radiosa Kütz.
Frequent throughout the island.
N. rhyncocephala Kütz.
Kings: Prospect Park, Brooklyn Water Supply.
N. varians Greg.
Kings: Prospect Park.
N. viridis Kütz.
Frequent throughout the island.

STAURONEIS Ehrb.
S. acuta W. Sm.
Kings: Coney Island, Woodman.
S. anceps Ehrb.
Kings: Prospect Park, Brooklyn Water Supply.
S. gracilis Ehrb.
Queens: Sag Harbor.
S. birostris Ehrb.
Queens: Arverne, Edwards.
S. aspera (Ehrb.) Kütz.
Queens: Arverne, Edwards.
S. Phoenicenteron Ehrb.
Kings: Coney Island, Woodman.

PLEUROSIGMA W. Sm.
P. angulatum W. Sm.
Queens: Rockaway Beach, Arverne, Edwards.
P. Balticum W. Sm.
Not infrequent in salt marshes throughout the island.
P. elongatum W. Sm.
Kings: Bay Ridge, Coney Island, Woodman.
P. Spencerii W. Sm.
Brooklyn Water Supply.

SCOLIOPLEURA Grun.
S. tumida (Bréb.) Rab.
Queens: Arverne, Edwards.

AMPHIPRORA Ehrb.
A. elegans W. Sm.
Queens: Arverne, Edwards.
A. navicularis Ehrb.
Queens: Arverne, Edwards.
A. ornata Bail.
Kings: Prospect Park, Brooklyn Water Supply.
A. pulchra Bail.
Queens: Arverne, Edwards.

GOMPHONEMA Ag.
  G. acuminatum Ehrb.
    Kings: Prospect Park, Brooklyn Water Supply.
  G. capitatum Ehrb.
    Kings: Prospect Park, Ridgewood.
  G. constrictum Ehrb.
    Kings: Prospect Park.

RHOICOSPHENIA Grun.
  R. curvata (Kütz.) Grun.
    Kings: Bay Ridge, Fort Hamilton.   *Queens:* Arverne, Edwards.

CYMBELLA Ag.
  C. Cistula Hempr.
    Kings: Prospect Park, Coney Island, Woodman.

ENCYONEMA Kütz.
  E. ventricosum Kütz.
    Kings: Prospect Park, Brooklyn Water Supply.

AMPHORA Ehrb.
  A. ovalis Kütz.
    Frequent throughout the island.

EPITHEMIA Bréb.
  E. gibba (Ehrb.) Kütz.
    Brooklyn Water Supply.
  E. Musculus Kütz.
    Queens: Arverne, Edwards.
  E. turgida (Ehrb.) Kütz.
    Kings: Coney Island, Woodman.   *Queens:* Arverne, Edwards.

NITZSCHIA Hass.
  N. acuminata W. Sm.
    Queens: Arverne, Edwards.
  N. fasciculata Grun.
    Kings: Bay Ridge.
  N. gracilis Hantzsch.
    Kings: Prospect Park.
  N. scalaris (Ehrb.) W. Sm.
    Kings: Coney Island, Woodman.
  N. Sigma W. Sm.
    Queens: Arverne, Edwards.
  N. sigmoidea (Ehrb.) W. Sm.
    Kings: Prospect Park.
  N. tryblionella H.
    Queens: Arverne, Edwards.
  N. tabellaria Grun.
    Kings: Prospect Park.

CYMATOPLEURA W. Sm.
   C. elliptica (Bréb.) W. Sm.
     Kings: Prospect Park.
   C. Salea (Bréb.) W. Sm.
     Kings: Prospect Park.
SURIRELLA Turp.
   S. elegans Ehrb.
     Not infrequent throughout the island.
   S. Febigerii Lewis.
     Queens: Arverne, Edwards.
   S. striatula Turp.
     Queens: Arverne, Edwards.
PYXILLA Grev.
   P. Balticum Grun.
     Queens: Arverne, Edwards.
DICLADIA Ehr.
   D. mitra Bail.
     Kings: Arverne, Edwards.

FUNGI.

SCHIZOMYCETES.

COCCACEAE.

STREPTOCOCCUS Billroth.
   S. pyogenes Rosenbach.
   S. erysipelatos Fehl.
     Common throughout.
MICROCOCCUS (Hallier) Cohn.
   M. pyogenes aureus (Passet) Rosenbach.
   M. pyogenes albus Rosenbach.
   M. gonorrhoeae (Neisser) Flügge.
   M. tetragonus Gaffky.
   M. ureae Pasteur.
   M. candicans Flügge.
     Frequent throughout the island.
SARCINA Goodsir.
   S. lutea Schröter.
   S. flava DeBary.
   S. aurantiaca Flügge.
     Common throughout in the air.

BACTERIACEAE.

BACTERIUM Ehr.
   B. Tuberculosis (Koch) Migula.
   B. influenzae Pfeiffer.

B. diptheritidis (Loefiler) Migula.
B. pneumonicum (Fried.) Migula.
B. aceti (Kütz.) Zopf.
B. acidi lactici (Hueppe) Migula.
All frequent throughout the island.

BACILLUS Cohn.
B. tetani Nicolaier.
B. typhi Gaffky.
B. coli (Esch.) Migula.
B. subtilis (Ehr.) Cohn.
Common throughout.

PSEUDOMONAS Migula.
P. pyocyanea (Gessard) Migula.
P. violaceus (Schröter) Migula.
P. fluorescens (Flügge) Migula.
Not infrequent.

CHLAMYDOBACTERIACAE.

CLADOTHRIX Cohn.
C. dichotoma Cohn.
In water.

BEGGIATOACEAE.

BEGGIATOA Trevisan.
B. mirabilis Cohn?
B. alba Trev.
Frequent.

EUMYCETES.

PHYCOMYCETES.

SYNCHYTRIACEAE.

SYNCHYTRIUM DC. & Woron.
S. anemones (DC.) Wor.
Kings: Brooklyn.

ALBUGINACEAE.

ALBUGO (Pers.) J. F. Gray.
A. cubicus (Pers.) Kuntze.
Kings: Brooklyn.

MUCORACEAE.

MUCOR Mich.
M. racemosus Fres.
Not infrequent.

M. mucedo L..
   Rare, Brooklyn.
CIRCINELLA Van Tiegh. & Le Mon.
   C. spinosa Van Tiegh.
      Brooklyn.
RHIZOPUS Ehr.
   R. stolonifer Ehr.
      Frequent.

ASCOMYCETES.

PROTOMYCETACEAE.

PROTOMYCES Unger.
   P. Linaria Fckl.
      Suffolk: Riverhead, Peck.

SACCHAROMYCETACEAE.

SACCHAROMYCES Meyen.
   S. cerevisiae Meyen.
      Frequent.
   S. niger Marp.
      Frequent.
   S. agglutinus Fr.
      Common.

GEOGLOSSACEAE.

GEOGLOSSUM Pers.
   G. glabrum Pers.
      Queens: Richmond Hill.   Cold Spring Harbor, Johnson.
LEOTIA Hill.
   L. lubrica (Scop.) Pers.
      Frequent throughout.
   L. chlorocephala Schw.
      Queens: Richmond Hill.   Cold Spring Harbor, Johnson.

HELVELLACEAE.

MORCHELLA Dill.
   M. esculenta (L.) Pers.
      Infrequent throughout the island.

HELOTIACEAE.

HELOTIUM Tode.
   H. citrinum Fr.
      Frequent throughout.

PHACIDIACEAE.

RHYTISMA Fr.
  R. acerinum (Pers.) Fr.
    Frequent throughout.
  R. Solidaginis (Schw.) Fr.
    Common throughout.

HYPODERMATACEAE.

LOPHODERMIUM Chev.
  L. arundinaceum (Schrad.) Chev.
    Suffolk: Wading River, Peck.

HYSTERIACEAE.

HYSTERIUM Tode.
  H. insidens Schw.
    Kings: Prospect Park.
  H. vulvatum Schw.
    Kings: New Lots.   *Queens:* Valley Stream.
  H. viticolum C. & P.
    Kings: New Lots.

ASPERGILLACEAE.

ASPERGILLUS Lk.
  A. glaucus (L.) Lk.
    Common throughout.
PENICILLIUM Lk.
  P. crustaceum (L.) Lk.
    Frequent throughout.

ARYSIBACEAE.

SPHAEROTHECA Lev.
  S. Castagnei Lev.
    Frequent throughout the island.
ERYSIBE Hedw.
  E. Cichoracearum DC.
    Kings: Prospect Park.
MICROSPHAERA Lev.
  M. Alni (DC.) Wint.
    Frequent throughout the island on Lilac.
UNCINULA Lev.
  U. Americana Howe.
    Kings: Flatbush, Zabriskie.
  U. necator Sw.
    Kings: Brooklyn.

HYPOMYCETEAE.

HYPOMYCES Fr.
   H. Lactifluorum Schw.
      Queens: Glen Cove.
NECTRIA Fr.
   N. cinnabarina (Tode) Fr.
PLEONECTRIA Winter.
   P. denigrata Winter.
      Kings: Bay Ridge.
HYPOCREA Fr.
   H. contorta (Schw.) Fr.
      Kings: Bay Ridge.   *Queens:* Valley Stream.
EPICHLOE Fr.
   E. Hypoxylon Pk.
      Queens: Rockaway Beach.
CORDYCEPS Fr.
   D. militaris (L.) Lk.
      Kings: Richmond Hill.

DOTHIDIACEAE.

PLOWRIGHTIA Sacc.
   P. morbosa (Schw.) Sacc.
      Frequent throughout on cherry and plum.
DOTHIDEA Fr.
   D. Graminis Pers.
      Queens: Rockaway Beach.
PHYLLACHORA Nitschke.
   P. Pteridis (Rab.) Fckl.
      Queens: Valley Stream.

CHAETOMIACEAE.

CHAETOMIUM Kunze.
   C. bostrychodes Zopf.
      Kings: Flatbush, Zabriskie.
CHAETOSPHAERIA Tul.
   C. longipila Pk.
      Kings: Flatbush, Zabriskie.
AMPHISPHAERIA Ces. & De Not.
   A. umbrina (Fr.) Wint.
      Kings: Ridgewood.
MASSARIA De Not.
   M. vomitoria B. & C.
      Kings: New Lots.   *Suffolk:* Saugerties, Peck.

HENDERSONIA Mont.
   H. epileuca B. & C.
     Suffolk: Saugerties, Peck.

VALSA Fr.
   V. ceratophora Tul.
     Queens: Valley Stream.

EUTYPA Tul.
   E. Acharii Tul.
     Queens: Valley Stream.

EUTYPELLA Cook.
   E. glandulosa (Ck.) Ell.
     Not infrequent on *Ailanthus*.

DIATRYPE Fr.
   D. stigma (Hoff.) Fr.
     Frequent throughout.

NUMMULARIA Tul.
   N. Bulliardi Tul.
     Frequent throughout.

USTULINA Tul.
   U. vulgaris Tul.
     Common throughout.

HYPOXYLON Bull.
   H. Howeanum Pk.
     Kings: New Lots.
   H. fuscum (Pers.) Fr.
     Frequent throughout.
   H. punctulatum B. & Rav.
     Kings: Flatbush.
   H. multiforme Fr.
     Queens: Valley Stream. Cold Spring Harbor, Johnson.
   H. coccineum Bull.
     Queens: Cold Spring Harbor, Johnson.
   H. effusum Nitz.
     Kings: Flatbush, Zabriskie.
   H. Sassafras Schw.
     Common throughout.

DALDINIA Ces. & De Not.
   D. concentrica (Balt.) Ces. & De Not.
     Frequent throughout.
   D. vernicosa (Schw.) Ces. & De Not.
     Queens: Rockaway Beach.

XYLARIA Fr.
  X. corniformis Fr.
    Kings: Prospect Park.
  X. polymorpha Grev.
    Queens: Cold Spring Harbor, Johnson.

HYPHOMYCETES.

MONILIA Pers.
  M. aurantiaca Peck & Sacc.
    Suffolk: Manor, Peck.
SPOROTRICHUM Lk.
  S. cinereum Pk.
    Suffolk: Manor, Peck.
ECHINOBOTRYS.
  E. atrium Cd.?
    Kings: Flatbush, Zabriskie.
STACHYOBOTRYS Corda.
  S. elongata Pk.
    Suffolk: Manor, Peck.
FUSICLADIUM Bon.
  F. fasciculatum C. & E.
    Suffolk: Manor, Peck.
CLADOSPORIUM Lk.
  C. epiphyllum Nees.
    Kings: Brooklyn.
  C. epimyces Cke.
    Kings: New Lots.
SEPTONEMA Cda.
  S. spilomeum Berk.
    Kings: Flatbush, Zabriskie.
CERCOSPORA Fr.
  C. Rhoinii C. & E.
    Suffolk: Manor, Peck.
  C. Eupatorii Pk.
    Suffolk: Wading River, Peck.
  C. Smilacis Thüm.
    Suffolk: Wading River, Peck.
  C. nymphaeacea C. & E.
    Suffolk: Riverhead, Peck.
SPORIDESMIUM Link.
  S. antiquum Cda.
    Kings: Flatbush, Zabriskie.
  S. epicoccoideum Pk.
    Kings: Brooklyn, New Lots.

MACROSPORIUM Fr.
  M. inquinans C. & E.
    Kings: Flatbush, Zabriskie.

ALTERNARIA Nees.
  A. tenuis Nees.
    Frequent throughout.

ISARIA Pers.
  I. Aranearum Schw.
    Suffolk: Manor, Peck.

SPOROCYBE Fr.
  S. cellare Pk.
    Kings: Flatbush, Zabriskie.

TUBERCULARIA Tode.
  T. vulgaris Tode.
    Common throughout the island.

FUSARIUM Desm.
  F. roseum Lk.
    Not infrequent.
  F. Scleradermatis Pk.
    Suffolk: Manor, Peck.

SPHAEROPSIDEAE.

PHYLLOSTICTA Pers.
  P. Negundinis Sacc. & Speg.
    Suffolk: Patchogue, Peck.
  P. Dioscoreae Cke.
    Suffolk: Riverhead, Peck.
  P. serotina Cke.
    Suffolk: Manor, Peck.
  P. Hibiscii Pk.
    Suffolk: Eastport and Patchogue.

PHOMA Fr.
  P. Candollei Sacc.
    Suffolk: Patchogue, Peck.

APOSPHAERIA Berk.
  A. aranea Pk.
    Kings: Flatbush, Zabriskie.

SPHAEROPSIS Mont.
  S. Smilacina Peck.
    Suffolk: Wading River, Peck.

DARLUCA Cast.
  D. Filum (Biv.) Cast.
    Not infrequent.

SEPTORIA Fr.
  S. Trichostematis Pk.
    Suffolk: Manor, Peck.
  S. Kalmicola B. & C.
    Queens: Cold Spring Harbor, Johnson.
SACIDIUM Nees.
  S. lignarium Pk.
    Kings: Flatbush, Zabriskie.
LEPTOSTROMA Fr.
  L. vulgare Fr.
    Not infrequent.
MELANCONIUM Lk.
  M. foliicolum Pk.
    Suffolk: Manor, Peck.

BASIDIOMYCETES.
USTILAGINACEAE.

USTILAGO Pers.
  U. austro-americanae Speg.
    Kings: Flatbush, Zabriskie.
  U. Montagnei major Desm.
    Suffolk: Wading River, Manor, Miller.
  U. Zeal (Berkm.) Magn.
    Common throughout the island.
  U. neglecta Niessl.
    Kings: Flatbush, Zabriskie.
  U. Rabenhorstiana Kühn.
    Kings: Flatbush, Zabriskie.
  U. utriculosa Tul.
    Kings: Flatbush, Zabriskie.
GRAPHIOLA Poiteau.
  G. Phoenicis (Moug.) Poit.
    On cultivated palms.

MELAMPSORACEAE.

COLEOSPORIUM Lev.
  C. Solidaginis (Schw.) Thum.
    Kings: Prospect Park.

PUCCINIACEAE.

GYMNOSPORANGIUM DC.
  G. fuscum Oerst = *G. Sabinae* (Dick.) Wint.
    Frequent throughout the island.

G. claviceps Cke. & Pk.   *Roestelia aurantiaca* Pk.
   Kings: Brooklyn.
G. globosum Farl.
   Queens: Glen Cove.
G. macropus Lk.
   Queens: Cold Spring Harbor, Johnson.

ROESTELIA Reb.
  R. Botryapitis Schw. *Gymnosporangium?*
   Suffolk: Riverhead, Peck.

UROMYCES Lev.
  U. Caladii Schw.
   Queens: Richmond Hill.
  U. gramineum Cke.
   Suffolk: Shelter Island, Clinton.
  U. Orobi Fuck.
   Kings: Flatbush, Zabriskie.
  U. Lespedezae Schw.
   Frequent throughout the island.
  U. Terebinthii (DC.) Winter.
   Kings: Flatbush, Zabriskie.
  U. Trifolii (Hedw.) Lév.
   Kings: Flatbush, Zabriskie.

PUCCINIA Pers.
  P. Anemones Pers.
   Kings: Brooklyn.
  P. Compositarum Schw.
   Queens: Glen Cove.
  P. Graminis Pers.
   Frequent throughout the island.
  P. Helianthi Schw.
   Kings: Flatbush, Zabriskie.
  P. Hieracii (Schum.) Mart.
   Queens: Glen Cove.
  P. mammillata Schraeb.
   Kings: Flatbush, Zabriskie.
  P. Sorghi Schw.
   Kings: New Lots.
  P. Xanthii Schw.
   Frequent throughout the island.

PHRAGMIDIUM Link.
  P. mucronatum Lk.
   Kings: Prospect Park.

AECIDIUM Pers.
- **A. Berberidis** Pers.
  Frequent throughout the island.
- **A. Caladii** Schw.
  Frequent throughout the island.
- **A. Lysimachiae** (Schlecht.) Wallr.
  Kings: Flatbush, Zabriskie.
- **A. Fraxinii** Schw.
  Queens: Cold Spring Harbor, Johnson.

UREDO Lev.
- **U. Chimaphilae** Peck.
  Suffolk: Amagansett, Peck.
- **U. luminata** Schw.
  Common throughout the island.
- **U. Potentillarum** Lk.
  Frequent throughout the island.

TREMELLACEAE.

TREMELLA L.
- **T. albida** Huds.
  Kings: Prospect Park.
- **T. foliacea** Pers.
  Queens: Valley Stream.
- **T. intumescens** Schw.?
  Queens: Richmond Hill.

EXIDIA Fr.
- **E. glandulosa** Fr.
  Queens: Cold Spring Harbor, Johnson.

EXOBASIDIACEAE.

EXOBASIDIUM Woron.
- **E. Vaccinii** Woron.
  Suffolk: Riverhead, Peck.

THELEPHORACEAE.

HYMENOCHAETE Lev.
- **H. rubiginosa** (Schrad.) Lev.
  Kings: Gravesend.   *Queens:* Cold Spring Harbor, Johnson.

STEREUM Pers.
- **S. complicatum** Fr.
  Frequent throughout the island.
- **S. Curtisii** Berk.
  Kings: Flatlands.

S. frustulosum (Pers.) Fr.
Kings: New Lots. *Queens:* Cold Spring Harbor, Johnson.
S. hirsutum (Willd.) Fr.
Kings: Flatlands, Prospect Park. *Queens:* Cold Spring Harbor, Johnson.
S. spadiceum Fr.
Queens: Valley Stream, Richmond Hill.
S. versicolor Fr.
Frequent throughout the island.

THELEPHORA Ehr.
T. pallida Schw.
Queens: Richmond Hill.
T. terrestris Ehr.
Queens: Rockaway Beach.

CRATERELLUS Fr.
C. cornucopioides Pers.
Queens: Cold Spring Harbor, Johnson.

CLAVARIACEAE.

CLAVARIA Vaill.
C. gracillima Peck.
Kings: Flatlands.
C. pyxidata Pers.
Queens: Richmond Hill.

HYDNACEAE.

HYDNUM L.
H. pallidum C. & E.
Suffolk: Manor, Peck.

IRPEX Fr.
I. cinnamomeus Fr.
Kings: New Lots, Flatlands, Richmond Hill.
I. paradoxus Fr.
Kings: Flatbush.
I. sinuosus Fr.
Kings: Flatbush.
I. Tulipiferae Schw.
Kings: Flatbush, Richmond Hill.

POLYPORACEAE.

MUCRONOPORUS E. & E.
M. gilvus (Schwein.) E. & E.
Kings: Flatlands.

POLYPORUS Fr.
  P. oblectans Berk.
    Queens: Richmond Hill.
  P. vulgaris Fr.?
    Kings: New Lots.   *Queens:* Richmond Hill.
  P. sulphureus Fr.
    Kings: Brooklyn.   *Queens:* Cold Spring Harbor, Johnson.
  P. chioneus Fr.
    Kings: Flatbush, Zabriskie.
  P. hispidus (Bull.) Fr.
    Suffolk: Quogue, Peck.
  P. pergamenus Fr.
    Not uncommon throughout the island.
  P. Caesarius Fr.
    Suffolk: Manor, Peck.
  P. betulinus (Bull.) Fr.
    Common throughout the island.
  P. hirsutus (Wulf.) Fr.
    Queens: Valley Stream.
  P. versicolor (L.) Fr.
    Common throughout the island.

TRAMETES Fr.
  T. Pini Fr.
    Suffolk: Eastport, Peck.
  T. sepium Berk.
    Not uncommon throughout the island.
  T. cinnabarina Jacq.
    Queens: Cold Spring Harbor, Johnson.
  T. rigida Berk.
    Kings: Prospect Park.

DAEDALEA Pers.
  D. Quercina (L.) Pers.
    Not infrequent throughout the island.
  D. unicolor (Bull.) Fr.
    Frequent throughout the island.

LENZITES Fr.
  L. Betulina (L.) Fr.
    Frequent throughout the island.
  L. sepiaria (Wulf.) Fr.
    Kings: Flatbush.
  L. bicolor Fr.
    Queens: Glen Cove.

FAVOLUS Pers.
  F. Europaeus Fr.
    Not infrequent throughout the island.

CYCLOMYCES Kütz.
   G. Greenii Berk.
      Queens: Cold Spring Harbor, Johnson.

FISTULINA Bull.
   F. Hepatica (Huds.) Fr.
      Frequent throughout the island on dead chestnut stumps.

BOLETUS L.
   B. castaneus Bull.
      Kings: Prospect Park, Richmond Hills. *Queens:* Cold Spring Harbor, Johnson.
   B. chrysenteron Bull.
      Kings: Prospect Park.
   B. Frostii Russ.
      Suffolk: Wading River, Peck.
   B. Americanus Pk.
      Queens: Cold Spring Harbor, Johnson.
   B. gracilis Pk.
      Queens: Cold Spring Harbor, Johnson.
   B. edulis Bull.
      Not infrequent.
   B. felleus Bull.
      Not uncommon throughout the island.
   B. scaber Bull.
      Queens: Glen Cove.

STROBILOMYCES Berk.
   S. strobilaceus (Scop.) Berk.
      Queens: Glen Cove. Cold Spring Harbor, Johnson.

AGARICACEAE.

SCHIZOPHYLLUM Fr.
   S. commune Fr.
      Frequent throughout the island.

PANUS Fr.
   P. stipticus (Bull.) Fr.
      Not uncommon throughout the island.
   P. laevis B. & C.
      Suffolk: Wading River, Peck.

MARASMIUS Fr.
   M. oreades (Bolt.) Fr.
      Frequent throughout the island.
   M. epiphyllus (Pers.) Fr.
      Queens: Valley Stream.

M. foetidus Fr.
   Manor, Peck.
M. albiceps Pk.
   Manor, Peck.

CANTHARELLUS Adans
   C. cinnabarinus Schw.
      Queens: Glen Cove.
   C. cibarius Fr
      Queens: Cold Spring Harbor, Johnson.

RUSSULA Fr.
   R. alutacea Fr.
      Queens: Glen Cove.
   R. brevipes Pk.
      Suffolk: Quogue, Peck.
   R. emetica Fr.
      Frequent throughout the island.
   R. foetans Fr.
      Kings: Flatlands.
   R. virescens Fr.
      Frequent throughout the island.

LACTARIUS Fr.
   L. piperitus Fr.
      Common throughout the island.
   L. vellereus Fr.
      Queens: Richmond Hill.
   L. volemus Fr.
      Queens: Glen Cove.
   L. alpinus Pk.
      Kings: Prospect Park, Richmond Hill.
   L. mutabilis Pk.
      Suffolk: Yaphank, Peck.

HYGROPHORUS Fr.
   H. miniatus Fr.
      Kings: Flatlands.

CORTINARIUS Fr.
   C. pulchrifolius Peck.
      Suffolk: Wading River, Peck.
   C. rubo-cinereus Peck.
      Suffolk: Wading River, Peck.

DERMOCYBE Fr.
   D. basalis Peck.
      Suffolk: Wading River, Peck.

COPRINUS Pers.
  C. comatus Fr.
    Kings: Prospect Park.
  C. atramentarius Fr.
    Kings: Prospect Park.   *Queens:* Glen Cove.
  C. micaceus Fr.
    Kings: Prospect Park.   *Queens:* Glen Cove.

HYPHOLOMA Fr.
  H. sublateritius Schaeff.
    Frequent throughout the island.

PSALLIOTA Fr.
  P. campestris L.
    Frequent throughout the island.

NAUCORIA Fr.
  N. Scirpicola Pk.
    Suffolk: Patchogue, Peck.

PHOLIOTA Fr.
  P. spectabilis Fr.
    L. I., Trask, N. Y. S. Rept. 34.

ENTOLOMA Fr.
  E. scabrinellus Peck.
    Suffolk: Wading River, Peck.

PLEUROTUS Fr.
  P. ostreatus Jacq.
    Frequent throughout the island.
  P. ulmarius Bull.
    Queens: Glen Cove.
  P. campanulatus Peck.
    Suffolk: Saugerties, Peck.

OMPHALIA Fr.
  O. subrufescens Pk.
    Queens: Glen Cove, Peck.

MYCENA Fr.
  M. galericulata Scop.
    Queens: Richmond Hill.

COLLYBIA Fr.
  C. velutipes Curtis.
    Kings: Prospect Park.   *Queens:* Rockaway Beach.

CLITOCYBE Fr.
  C. laccatus Scop.
    Kings: Prospect Park, Richmond Hill.
  C. trullissatus Ell.
    Long Island, Peck.

C. solitarius Bull.
Suffolk: Wading River, Peck.
C. rhagadiosus Fr.
Suffolk: Wading River, Peck.
TRICHOLOMA Fr.
T. sejunctum Low.
Queens: Manor, Peck. *Suffolk:* Quoque, Peck.
T. grave Peck.
Manor, Peck.
ARMILLARIA Fr.
A. melleus Vahl.
Frequent throughout the island.
AMANITA Fr.
A. vernus Bull. (*A. phalloides*).
Queens: Glen Cove.
A. muscarius L.
Queens: Roslyn, Glen Cove.
A. rubescens Fr.
Queens: Glen Cove.

GASTEROMYCETES.

LYCOPERDACEAE.

LYCOPERDON L.
L. gemmatum Fr.
Kings: Flatlands.
L. pyriforme Schaeff.
Common throughout the island.
GEASTER L.
G. hygrometricus ? (L.) Pers.
Common throughout the island.

NIDULARIACEAE.

CYATHUS Haller.
C. striatus (Huds.) Hoff.
Kings: Flatbush.
CRUCIBULUM Tul.
C. vulgare Tul.
Frequent throughout the island.

SCLERODERMATACEAE.

SCLERODERMA Pers.
S. Bovista Fr.
Kings: Prospect Park, Flatlands.

S. Geaster Fr.
   Suffolk: Manor, Peck.
S. vulgare Fr.
   Frequent throughout the island.

PHALLACEAE.

PHALLUS Mech.
   P. impudicus L.
   Infrequent throughout island.
   P. duplicatus Bosc.
   Kings: Brooklyn.
MUTINUS.
   M. bovinus Morgan.
   Queens: Glen Cove. *Suffolk:* Manor, Peck.

LICHENES.

CLADONIACEAE.

BAEOMYCES Pers.
   B. roseus Pers.
   Queens: Lawrence, Valley Stream.
CLADONIA Hoffm.
   C. pyxidata (L.) Fr.
   Frequent throughout the island.
   C. gracilis verticillata Fr.
   Frequent throughout the island.
   C. Papillaria (Ehrh.) Hoffm.
   Suffolk: Orient, Young.
   C. furcata (Huds.) Fr.
   Frequent throughout the island.
   C. rangiferina (L.) Hoffm.
   Frequent throughout the island.
   C. uncialis (L.) Fr.
   Kings: Ridgewood, Brainerd.
   C. cornucopioides (L.) Fr.
   Queens: Richmond Hill.
   C. cristatella Tuck.
   Frequent throughout the island.
   C. bacillaris. (?)
   Kings: New Lots, Brainerd.

LECIDEACEAE.

BIATORA Fr.
  B. fusco-rubella Hoffm.
    Queens: Valley Stream.

LECIDEA Ach.
  L. albocoerulescens Fr.
    Queens: Richmond Hill.

BUELLIA De Not.
  B. parasema (Ach.) Th. Fr.
    Frequent throughout the island.

GRAPHIDACEAE.

OPEGRAPHA Humb.
  O. varia (Pers.) Fr.
    Queens: Richmond Hill.

GRAPHIS Ach.
  G. scripta Ach.
    Frequent throughout the island.

PHYSCIACEAE.

PLACODIUM Ach.
  P. vittellinum (Ebr.) Naeg. & Hepp.
    Kings: Bay Ridge.

PYXINE Fr.
  P. sorediata Fr.
    Kings: Gowanus, Brainerd.

PHYSCIA DC.
  P. aquila (Ach.) Nyl.
    Kings: New Lots, Brainerd.
  P. stellaris (L.) Tuck.
    Frequent throughout the island.
  P. tribacia (Ach.) Tuck.
    Kings: New Lots, Brainerd. *Queens:* Richmond Hill, Glen Cove.

THELOSCHISTES Norm.
  T. parietinus (L.) Norm.
    Kings: Flatbush. *Queens:* Flushing, Brainerd. *Suffolk:* Orient, Young.
  T. lychneus Nyl.
    Kings: Flatbush.
  T. chrysophthalmus (L.) Norm.
    Frequent throughout the island.

PARMELIACEAE.

LECANORA Ach.
- **L. pallida** (Schreb.) Schaer.
  Kings: Flatbush, Ridgewood, Brainerd. *Queens:* Valley Stream.
- **L. subfusca** (L.) Ach.
  Queens: Richmond Hill, Valley Stream.
- **L. varia** (Ehrh.) Nyl.
  Kings: Ridgewood, Brainerd. *Queens:* Glen Cove.

PARMELIA Ach.
- **P. crinita** Ach.
  Kings: Ridgewood, Brainerd.
- **P. perforata** (Jacq.) Ach.
  Queens: Valley Stream.
- **P. perlata** (L.) Ach.
  Kings: Ridgewood, Brainerd.
- **P. tiliacea** (Hoffm.) Floerk.
  Queens: Valley Stream.
- **P. Borreri** Ach.
  Kings: Gowanus, Brainerd; Flatlands. *Queens:* Glen Cove, *Suffolk:* Orient, Young.
- **P. saxatilis** (L.) Fr.
  Kings: Ridgewood, Brainerd. *Queens:* Valley Stream, *Suffolk:* Orient, Young.
- **P. caperata** (L.) Ach.
  Common throughout the island.
- **P. conspersa** (Ehrh.) Ach.
  Kings: Gowanus, Brainerd; Prospect Park. *Queens:* Glen Cove.
- **P. olivacea** (L.) Ach.
  Queens: Valley Stream.
- **P. rutilans** (?)
  Kings: Flatbush, Brainerd.

CETRARIA Ach.
- **C. Islandica** (L.) Ach.
  Queens: Richmond Hill.
- **C. ciliaris** (Ach.) Tuck.
  Kings: East New York, Brainerd.
- **C. lacunosa** Ach.
  Queens: Jamaica, Brainerd.

RAMALINA Ach.
- **R. calicaris** (L.) Fr.
  Frequent throughout the island.

ALECTORIA Ach.
  A. jubata (L.) Tuck.
    Queens: Jamaica, Brainerd.

USNEA Ach.
  U. barbata (L.) Fr.
    Frequent throughout the island.
  U. angulata Ach.
    Suffolk: Montauk Point, Brainerd.
  U. trichodea Ach.
    Suffolk: Orient, Young.

VERRUCARIACEAE.

CONOTREMA Tuck.
  C. urceolatum (Ach.) Tuck.
    Queens: Valley Stream, Jelliffe.

PERTUSSARIA DC.
  P. velata (Turn.) Nyl.
    Kings: Ridgewood, Brainerd.
  P. communis DC.
    Kings: Ridgewood, Brainerd. *Queens:* Richmond Hill.

ENDOCARPACEAE.

ENDOCARPON Hedw.
  E. miniatum (L.) Schaer.
    Suffolk: Orient, Young.

COLLEMACEAE.

COLLEMA Hoffm.
  C. flaccidum Ach.
    Queens: Jamaica, Brainerd, Valley Stream.

LEPTOGIUM Fr.
  L. tremelloides (L. f.) Fr.
    Queens: Jamaica, Brainerd, Valley Stream.

PANNARIACEAE.

PELTIGERA Willd.
  P. aphthosa (L.) Hoffm.
    Kings: New Lots, Brainerd.
  P. canina (L.) Hoffm.
    Queens: Jamaica, Brainerd.

STICTA Schreb.
  S. pulmonaria (L.) Ach.
    Frequent throughout the island.
  S. amplissima (Scop.) Mass.
    Kings: New Lots, Brainerd.

HEPATICAE.

RICCIACEAE.

RICCIA L.
  R. fluitans L.
    Kings: Canarsie, Brainerd.
      forma terrestris.
    Queens: Freeport, Howe.
  R. natans L.
    Kings: Gowanus, Brainerd; Ridgewood.

MARCHANTIACEAE.

CONOCEPHALUM Wiggers.
  C. conicum (L.) Dumort.
    Common throughout.
MARCHANTIA L.
  M. polymorpha L.
    Frequent throughout the island.

METZGERIACEAE.

RICCARDIA S. F. Gray.
  R. pinguis (L.) S. F. Gray.
    Kings: Ridgwood, Brainerd.
  R. major (Nees) Lindb.
    Queens: Freeport, Howe.
PELLIA Raddi.
  P. epiphylla (L.) Nees.
    Queens: Jamaica, Brainerd.
PALLAVICINIA S. F. Gray.
  P. Lyelli (Hook.) S. F. Gray.
    Queens: Freeport, Howe.

JUNGERMANNIACEAE.

PLAGIOCHILA Dumort.
  P. porelloides Lindenb.
    Queens: Roslyn, Brainerd.

LOPHOCLEA Dumort.
   L. heterophylla (Schrad.) Dumort.
     Queens: Freeport, Howe.

CHILOSCYPHUS Corda.
   C. polyanthus (L.) Corda; var. **rivularis** Nees.
     Queens: Freeport, Howe.

SACCOGYNA Dumort.
   S. graveolens (Schrad.) Lindb.
     Queens: Freeport, Howe.

CEPHALOZIA Dumort.
   C. catenulata (Hüben.) Lindb.
     Queens: Freeport, Howe.
   C. connivens (Dicks.) Lindb.
     Queens: Freeport, Howe.

LEPIDOZIA Dumort.
   L. setacea (Web.) Mitt.
     Queens: Freeport, Howe.

ODONTOSCHISMA Dumort.
   O. sphagna (Dicks.) Dumort.
     Queens: Freeport, Howe.

KANTIA S. F. Gray.
   K. Trichomanis (L.) S. F. Gray.
     Queens: Freeport, Howe.
   K. Sullivantii (Austin) Underwood.
     Queens: Freeport, Howe.

BLEPHAROSTOMA Dumort.
   B. nematodes (Gottsche) Underwood and Cooke.
     Queens: Freeport, Howe.

SCAPANIA Dumort.
   S. nemorosa (L.) Nees.
     Kings: East New York, Brainerd.

TRICHOCOLEA Dumort.
   T. tomentella (Ehr.) Dumort.
     Kings: New Lots, Brainerd.

RADULA L.
   R. complanata (L.) Dumort.
     Kings: New Utrecht, Brainerd.

PORELLA L.
   B. platyphylla (L.) Lindb.
     Frequent throughout the island.

FRULLANIA Raddi.
   F. Eboracensis Gottsche.
     Not infrequent on trees.
   F. Asa-Grayana Mont.
     Kings: Ridgewood, Brainerd.

ANTHOCERATACEAE.

ANTHOCERAS L.
  A. laevis L.
    Queens: Freeport, Howe.

NOTOTHYLAS Sulliv.
  N. orbicularis (Schwein) Sulliv.
    Queens: Freeport, Howe.

MUSCI.

SPHAGNACEAE.

SPHAGNUM L.
  S. acutifolium Ehr.
    Not infrequent.
  S. cuspidatum Ehr.
    Kings: New Utrecht, Brainerd.
  S. cymbifolium Ehr.
    Not infrequent.
  S. molle Sull.
    Suffolk: Northport, M. L. Saniel.
  S. subsecundum Nees.
    Kings: New Utrecht, New Lots, Brainerd.
  S. subsecundum obesum (Wils.) Schimp.
    Kings: Greenwood, Brainerd.

PHASCACEAE.

ACAULON C. Müll.
  A. muticum (Schreb.) C. Müll.
    Kings: Brooklyn, Austin.

BRUCHIACEAE.

PLEURIDIUM Brid.
  P. alternifolium (Dicks.) Rab.
    Queens: Glen Cole, M. L. Saniel.

WEISIACEAE.

HYMENOSTYLIUM Brid.
  H. curvirostre (Ehrh.) Lindb.
    Kings: Bath, C. H. Hall.

WEISIA Hedw.
  W. viridula (L.) Hedw.
    Common throughout.

DICRANACEAE.

DICRANELLA Schimp.
  D. varia (Hedw.) Schimp.
    Suffolk: Northport, M. L. Saniel.
  D. heteromalla (L.) Schimp.
    Frequent throughout.

DICRANUM Hedw.
  D. spurium Hedw.
    Queens: Huntington, J. E. Rogers.
  D. undulatum Ehr.
    Suffolk: Wading River, E. S. Miller.
  D. scoparium (L.) Hedw.
    Frequent.
  D. montanum Hedw.
    Suffolk: Northport, M. L. Saniel.
  D. flagellare Hedw.
    Not infrequent.
  D. fulvum Hook.
    Queens: Flushing, Brainerd.

LEUCOBRYACEAE.

LEUCOBRYUM Hampe.
  L. glaucum (L.) Schimp.
    Not infrequent throughout.

FISSIDENTACEAE.

FISSIDENS Hedw.
  F. osmundoides (Sw.) Hedw.
    Kings: College Point, Brainerd.
  F. taxifolius (L.) Hedw.
    Not infrequent.

DITRICHACEAE.

CERATODON Brid.
  C. purpureus (L.) Brid.
    Frequent throughout.
  C. purpureus aristatus Aust.
    Suffolk: Orient, Young.

DITRICHUM Timm.
  D. tortile (Schrad.) Lindb.
    Kings: Evergreen Cemetery, Brainerd.   *Queens:* Fresh Pond, Hulst; Huntington, M. L. Saniel.

D. vaginans (Sull.) Hampe.
  Queens: Fresh Pond, Hulst.
D. pallidum (Schreb.) Hampe.
  Not infrequent throughout.

POTTIACEAE.

POTTIA Ehr.
  P. truncatula (L.) Lindb.
    Infrequent in roadsides.

BARBULA Hedw.
  B. unguiculata (Huds.) Hedw.
    Kings: Brooklyn. *Suffolk:* Smithtown, M. L. Saniel.

GRIMMIACEAE.

GRIMMIA Ehr.
  G. Pennsylvanica Schwaeg.
    Kings: Gowanus, Brainerd. *Suffolk:* M. L. Saniel.

RACOMITRUM Brid.
  R. aciculare (L.) Brid.
    Kings: Gravesend, Bath, Brainerd.

HEDWIGIA Ehr.
  H. albicans (Web.) Lindb.
    Kings: New Lots, Brainerd. *Queens:* Glen Cove. *Suffolk:* Northport, M. L. Saniel.

ORTHOTRICHACEAE.

ULOTA Mohr.
  U. Americana (P. B.) Lindb.
    Queens: Richmond Hill.
  U. crispa (L.) Brid.
    Queens: Jamaica, Brainerd. *Suffolk:* E. Hampton, Anna M. Vail.

ORTHOTRICHUM Hedw.
  O. cupulatum Hoff.
    Suffolk: E. Hampton, Anna M. Vail.
  O. Braunii Bry. Eur.
    Suffolk: Centreport, M. L. Saniel.
  O. pusillum Mitt.
    Queens: Richmond Hill.
  O. psilocarpum Jas.
    Queens: Richmond Hill, Flatbush.

DRUMMONDIA Hook.
  D. clavellata Hook.
    Kings: Greenwood, Brainerd. *Queens:* Jamaica, Brainerd; Lloyd's Neck, M. L. Saniel; Richmond Hill.

GEORGIACEAE.

GEORGIA Ehrh.
  G. pellucida (L.) Raben.
    Kings: New Lots, Brainerd. *Queens:* Richmond Hill.

FUNARIACEAE.

PHYSCOMITRIUM Brid.
  P. pyriforme (L.) Brid.
    Kings: Brooklyn, New Lots, Gowanus, Brainerd. *Queens:* Glen Cove, M. L. Saniel.

FUNARIA Schreb.
  F. flavicans Mx.
    Kings: Flatbush, Brainerd.
  F. hygrometrica (L.) Sibth.
    Not infrequent.

BRYACEAE.

LEPTOBRYUM Schimp.
  L. pyriforme (L.) Schimp.
    Kings: Flatbush. *Queens:* Lloyd's Neck, M. L. Saniel; Richmond Hill.

BRYUM Dill.
  B. argenteum L.
    Common throughout.
  B. caespiticium L.
    Kings: Greenwood, Brainerd; Prospect Park. *Queens:* Richmond Hill. *Suffolk:* Northport, M. L. Saniel.
  B. intermedium (Lud.) Brid.
    Kings: New Lots, East New York, Brainerd.

RHODOBRYUM Schimp.
  R. roseum (Weis) Lind.
    Kings: Ridgewood, Brainerd. *Suffolk:* Pardegot, Brainerd.

MNIACEAE.

MNIUM Dill.
  M. hornum L.
    Kings: New Lots, Brainerd, Jelliffe. *Queens:* Valley Stream.
  M cuspidatum (L.) Leyss.
    Common throughout.

M. affine Bland.
Frequent throughout.
M. punctatum (L., Schreb.) Hedw.
Kings: New Lots, Brainerd.

AULACOMNIACEAE.

AULACOMNIUM Schwägr.
A. heterostichum B & S.
Kings: New Lots, Brainerd. *Suffolk:* The Cove, M. L. Saniel.
A. palustre (L.) Schw.
Kings: New Lots, Brainerd. *Suffolk:* M. L. Saniel.

BARTRAMIACEAE.

BARTRAMIA Hedw.
B. pomiformis (L.) Hedw.
Kings: Pike. *Queens:* Cold Spring; M. L. Saniel, Rosyln, Brainerd.

PHILONOTUS Brid.
P. fontana (L.) Brid.
Queens: Huntington, M. L. Saniel. *Suffolk:* Northport, M. L. Saniel.

POLYTRICHACEAE.

CATHARINEA Ehrh.
C. undulata (L.) Web. & Mohr.
Frequent throughout.
C. angustata Brid.
Frequent throughout.
C. crispa (Aust.) James.
Kings: Prospect Park, Brainerd; Flatbush Water Works.

POGONATUM Beauv.
P. brevicaule Beauv.
Frequent throughout.

POLYTRICHUM L.
P. piliferum Schreb.
Suffolk: East Meadow, M. L. Saniel.
P. Juniperinum Willd.
Infrequent.
P. commune L.
Common throughout.
P. Ohioense R. & C.
Frequent throughout.

BUXBAUMIACEAE.

DIPHYSCIUM (Ehrh.) Mohr.
  D. sessile (Schmid.) Lindb.
    Suffolk: Centreport, M. L. Saniel.

PLEUROCARPAE.

FONTINALACEAE.

FONTINALIS L.
  F. antipyretica L.
    Queens: Flushing, Pike; Valley Stream. *Suffolk:* Moriches, Brainerd.

DICHELYMA Myrin.
  D. capillaceum Schimp.
    Kings: Ridgewood, Gowanus, Brainerd. *Suffolk:* Northport, M. L. Saniel.

CRYPHAEACEAE.

LEUCODON Schwäger.
  L. julaceus (Hedw.) Sull.
    Kings: Bath, Brainerd. *Suffolk:* Moriches, Brainerd.

LESKEACEAE.

ANOMODON (Hook. Tayl.)
  A. attenuatus (Schreb.) Hüben.
    Kings: New Lots, Flatbush, Brainerd. *Queens:* Valley Stream, Richmond Hill.
  A. rostratus (Hedw.) Schimp.
    Kings: Brooklyn, Gowanus, Ridgewood, Brainerd.

THUIDIUM Bry. Eur.
  T. delicatulum (L.) Mitt.
    Frequent throughout.
  T. recognitum Hedw.
    Queens: Richmond Hill. *Suffolk:* Northport, M. L. Saniel.
  T. Virginianum Lindb.
    Queens: Lloyd's Neck, M. L. Saniel.

HYPNACEAE.

PLATYGYRIUM Bry. Eur.
  P. repens (Brid.) Bry. Eur.
    Suffolk: Smithtown, M. L. Saniel.

PYLAISIA Br. Sch.
  P. sub-denticulata Sch.
    Queens: Richmond Hill.
  P. velutina B. & S.
    Kings: New Lots, Brainerd; Cypress Hills, Hulst.

CYLINDROTHECIUM Bry. Eur.
  C. cladorrhizans (Hedw.) Sch.
    Suffolk: Northport, M. L. Saniel.
  C. seductrix (Hedw.) Sull.
    Kings: Gravesend, New Utrecht, Brainerd; Prospect Park.
    Queens: Richmond Hill.

CLIMACIUM Web. & Mohr.
  C. Americanum Brid.
    Not infrequent throughout.

BRACHYTHECIUM Bry. Eur.
  B. salebrosum (Hoffm.) Bry. Eur.
    Queens: Richmond Hill.
  B. rutabulum (L.) Bry. Eur.
    Suffolk: Northport, M. L. Saniel.
  B. plumosum (Sch.) Bry. Eur.
    Frequent throughout.
  B. acuminatum (Beauv.) Bry. Eur.
    Suffolk: Centreport, M. L. Saniel.
  B. rivulare Bry. Eur.
    Kings: Gowanus, New Lots, Brainerd. *Suffolk:* Centreport, M. L. Saniel.
  B. velutinum (L.) Bry. Eur.
    Kings: New Lots, Brainerd.

EURYNCHIUM Bry. Eur.
  E. strigosum (Hoff.) Bry. Eur.
    Kings: Ridgewood, New Utrecht, Brainerd. *Suffolk:* Northport, M. L. Saniel.
  E. hians (Hedw.) Jäger & Sauerb.
    Kings: New Utrecht, Brainerd. *Suffolk:* Northport, M. L. Saniel.
  E. Boscii (Schw.) B. &. S.
    Frequent throughout.

RHAPHIDOSTEGIUM (Bry. Eur.) De Not.
  R. demissum (Nils. & Sch.) De Not.
    Queens: Valley Stream.

PLAGIOTHECIUM Bry. Eur.
  P. sylvaticum (Huds.) Br. & Sch.
    Suffolk: Centreport, M. L. Saniel.

P. Sullivantiae Sch.
 Kings: Gowanus, Brainerd.
P. denticulatum (L.) Br. & Sch.
 Kings: Canarsie, Brainerd.
P. turfaceum (Lind.) Limp.
 Suffolk: Northport, M. L. Saniel.

AMBLYSTEGIUM Schimp.
 A. adnatum (Hedw.) Br. & Sch.
  Kings: New Lots, Brainerd. *Suffolk:* Northport, M. L. Saniel.
 A. serpens (L.) Br. & Sch.
  Kings: Prospect Park ; New Lots, Bath, Brainerd.
 A. serpens orthocladon (Aust.). ?
  Kings: New Lots, Bath, Brainerd.
 A. riparium (L.) Br. & Sch.
  Frequent throughout.
 A. varium (Hedw.) Lindb.
  Kings: New Lots, Brainerd.

HYPNUM L.
 H. hispidulum Brid.
  Suffolk: Centreport, M. L. Saniel.
 H. paludosum Sulliv.
  Suffolk: Centreport, M. L. Saniel.
 H. Sullivantii Spruce.
  Kings: Bath, Brainerd.
 H. serrulatum Hedw.
  Frequent throughout.

HYLOCOMIUM Br. & Sch.
 H. triquetrum (L.) Br. &. Sch.
  Kings: New Lots, Brainerd. *Suffolk:* Northport, M. L. Saniel.
 H. parietinum (L.) Lindb.
  Kings: New Lots, Bath, Brainerd.
 H. proliferum (L.) Lindb.
  Suffolk: Northport, M. L. Saniel.

STEREODON Mitt.
 S. cupressiforme (L.) Brid. ?
  Not infrequent.
 S. pallescens (Hedw.) Lindb.
  Kings: New Lots, Brainerd.
 S. imponens (Hedw.) Brid.
  Frequent throughout.
 S. Haldianum (Grev.) Lindb.
  Frequent throughout.

THELIA Sulliv.
 T. hirtella (Hedw.) Sulliv.
  Frequent throughout.

PTERIDOPHYTA.

OPHIOGLOSSACEAE.

OPHIOGLOSSUM L.
  O. vulgatum L. f. 1.*
  Long Island, Hulst. ?

BOTRYCHIUM Sw.
  B. simplex Hitch. f. 2.
  Suffolk: Miller and Young.
  B. matricariaefolium A. Br. f. 4.
  Long Island, Hulst.
  B. ternatum dissectum (Sp.) Milde. f. 5.
  Queens: Richmond Hill, Hulst. *Suffolk:* Miller and Young.
  Var. **obliquum** (Muhl.) Milde.
  Queens: Richmond Hill. *Suffolk:* Miller and Young.
  B. Virginianium (L.) Sw. f. 7.
  Frequent throughout the island.

OSMUNDACEAE.

OSMUNDA L.
  O. regalis L. Royal Fern. f. 8.
  Frequent throughout the island.
  O. cinnamomea L. Cinnamon Fern. f. 9.
  Common throughout the island.
  O. Claytoniana L. f. 10.
  Frequent throughout the island.

POLYPODIACEAE.

ONOCLEA L.
  O. sensibilis L. Sensitive Fern. f. 14.
  Common throughout the island.

WOODSIA R. Br.
  W. Ilvensis (L.) R. Br. f. 16.
  Long Island, W. Dunham.
  W. obtusa (Spreng.) Torr. Woodsia. f 21.
  Suffolk: Greenport, 1861, Herb. L. I. Historical Society (H. S.).

DICKSONIA L'Her.
  D. punctilobula (Mx.) Gray. Dicksonia. f. 22.
  Kings: Cypress Hills, Hulst; New Lots, Brainerd. *Suffolk:* Miller and Young.

* Figure numbers refer to the figures in Britton and Brown's " Illustrated Flora."

DRYOPTERIS Adans. (*Aspidium*).
   D. acrostichoides (Mx.) Kuntze. f. 27.
     Frequent throughout the island.
   D. Schweinitzii (Beck) B. S. P.
     Suffolk: East Hampton, Mrs. L. D. Pychowska.
   D. Noveboracensis (L.) A. Gray. f. 29.
     Frequent throughout the island.
   D. Thelypteris (L.) A. Gray. f. 30.
     Frequent throughout the island.
   D. cristata (L.) A. Gray. f. 33.
     Suffolk: Miller & Young.
   D. marginatus (L.) A. Gray. f. 35.
     Common throughout the island.
   D. spinulosa (Retz.) Kuntze. f. 37.
     Common throughout the island.
   D. spinulosa intermedia (Muhl.) Underw.
     Frequent throughout the island.

PHEGOPTERIS Fee.
   P. Phegopteris (L.) Underw. f. 39.
     Kings: Prospect Park; Flatbush, Brainerd.
   P. hexagonoptera (Mx.) Fee. f. 40.
     Kings: Calverley, Cypress Hills, Hulst. *Queens:* Richmond Hill, Hulst, J.
   P. Dryopteris (L.) Fee. Beach-fern. f. 41.
     Queens: Richmond Hill.

WOODWARDIA J. E. Smith.
   W. Virginica (L.) Smith. Chain-fern. f. 42.
     Queens: Calverley, Fresh Pond, Hulst. *Suffolk:* Miller & Young.
   W. areolata (L.) Moore. f. 43.
     Kings: Forbell's Landing, Hulst. *Queens:* Fresh Pond, Hulst, Jamaica, Brainerd; Valley Stream. *Suffolk:* Greenport, Tillinghast, Miller & Young.

ASPLENIUM L.
   A. platyneuron (L.) Oakes. f. 49.
     Frequent throughout the island.
   A. Trichomanes L. f. 50.
     Long Island, Hulst. *Queens:* Astoria, Herb. H. S.
   A. acrostichoides Sw. Shield Fern. f. 57.
     Frequent throughout the island.
   A. Filix-foemina (L.) Bernh. f. 58.
     Frequent throughout the island.

ADIANTUM L.
   A. pedatum L. Maiden-hair Fern. f. 60.
     Occasional throughout the island.

PTERIS L.
  P. aquilina L. Brake. f. 61.
    Frequent throughout the island.

POLYPODIUM L.
  P. vulgare L. Polybody. f. 71.
    Common throughout the island.

EQUISETACEAE.

EQUISETUM L.
  E. arvense L. Common Horsetail. f. 77.
    Common throughout the island.
  E. fluviatile L. f. 82.
    Queens: Woodside, Hulst. *Suffolk:* Miller & Young.

LYCOPODIACEAE.

LYCOPODIUM L.
  L. inundatum L. f. 90.
    Suffolk: Miller & Young.
  L. alopecuroides L. Club Moss. f. 91.
    Kings: Forbell's Landing, Hulst. *Queens:* Calverley. *Suffolk:* East Hampton, Mrs. L. D. Pychowska.
  L. obscurum L. Ground Pine. f. 92.
    Long Island, Hulst. *Kings:* Calverley. *Suffolk:* Miller & Young.
  L. clavatum L. f. 96.
    Long Island, Hulst. *Suffolk:* Miller & Young.
  L. complanatum L. Trailing Christmas Green. f. 98.
    Queens: Richmond Hill. *Suffolk:* Miller & Young.

SELAGINELLACEAE.

SELAGINELLA Beauv.
  S. apus (L.) Spreng. Selaginella. f. 101.
    Queens: Calverley. *Suffolk:* Miller & Young.

SPERMATOPHYTA.

GYMNOSPERMAE.

PINACEAE.

PINUS L.
  P. Strobus L. White Pine. f. 110.
    Frequent throughout the island.
  P. Virginiana Mill. Hemlock Pine. f. 115.
    Long Island, Hulst. *Suffolk:* Miller & Young.

P. echinata Mill. Spruce Pine. f. 116.
  Kings: Ridgewood, Brainerd. Long Island, Hulst. *Suffolk:* Miller & Young.
P. rigida Mill. Pitch Pine. f. 119.
  Frequent throughout the island.

LARIX Adans.
  L. laricina (Du Roi) Koch. Larch. f. 120.
    Long Island, Hulst. *Kings:* Prospect Park. *Queens:* Richmond Hill.

PICEA Link.
  P. Canadensis (Mill.) B.S.P. Spruce. f. 121.
    Long Island, Hulst.
  P. Mariana (Mill.) B.S.P. Black Spruce. f. 122.
    Suffolk: Miller & Young.

TSUGA Carr.
  T. Canadensis (L.) Carr. Hemlock Spruce. f. 124.
    Long Island, Hulst. *Kings:* Prospect Park. *Suffolk:* Miller & Young.

ABIES Juss.
  A. balsamea (L.) Mill. Balsam. f. 126.
    Long Island, Hulst.?

THUJA L.
  T. occidentalis L. White Cedar. f. 129.
    Frequent throughout the island, planted.

CHAMAECYPARIS Spach.
  C. thyoides (L.) B.S.P. f. 130.
    Suffolk: Miller & Young.

JUNIPERUS L.
  J. communis L. Juniper. f. 131.
    Frequent throughout the island.
  J. Virginiana L. f. 133.
    Common throughout the island.

TAXACEAE.

TAXUS L.
  T. minor (Mx) Britton. f. 135.
    Kings: Prospect Park.

ANGIOSPERMAE.

MONOCOTYLEDONES.

TYPHACEAE.
TYPHA L.
  T. latifolia L. Cat-tail. f. 136.
    Frequent throughout the island.
  T. angustifolia L. f. 137.
    Frequent throughout the island.

SPARGANIACEAE.
SPARGANIUM L.
  S. eurycarpum Eng. Bur-reed. f. 138.
    Kings: Forbell's, Hulst. *Suffolk:* Miller & Young.
  S. androcladum (Eng.) Morong. f. 139.
    Queens: Cold Spring, Hulst; Flushing, Hempstead, N. Hempstead, Oyster Bay, Bisky. *Suffolk:* Miller & Young.
  S. simplex Huds. f. 140.
    Queens: Richmond Hill; Flushing, Bisky; Glen Cove. *Suffolk:* Greenport, Tillinghast ; Miller & Young.

NAIADACEAE.
POTAMOGETON L.
  P. natans L. Pond-weed. f. 142.
    Queens: Calverley; Flushing, Hempstead, Bisky.
  P. Oakesianus Robbins. f. 143.
    Queens: Cold Spring, Hulst ; Morong.
  P. amplifolius Tuck. f. 144.
    Kings: Cypress Hills, Hulst.
  P. Nuttallii Ch. & Sch. f. 146.
    Long Island, Hulst. *Queens:* Bisky. *Suffolk:* Miller & Young.
  P. heterophyllus Schreb. f. 150.
    Queens: Jamaica, Leggett.
  P. foliosus Raf. f. 162.
    Suffolk: Miller & Young.
  P. pusillus L. f. 168
    Queens: N. Hempstead, Leggett. *Suffolk:* Miller & Young.
  P. diversifolius Raf. f. 170.
    Suffolk: Miller & Young.
  P. pectinatus L. f. 173.
    Suffolk: Miller & Young.
  P. Robbinsii Oakes. f. 175.
    Kings: Forbell's Landing, Hulst. *Suffolk:* Miller & Young.

RUPPIA L.
   R. maritima L. Ditch-grass. f. 176.
     Frequent along the shores.

ZANNICHELLIA L.
   Z. palustris pedunculata Gray. Horned Pond Weed. f. 178.
     Queens: Flushing, Bisky.

NAIAS L.
   N. flexilis (Willd.) R. & S. Naiad. f. 180.
     Queens: Flushing, Leggett. *Suffolk:* Miller & Young.

ZOSTERA L.
   Z. marina L. Eel-grass. f. 183.
     Common along the shores.

SCHEUCHZERIACEAE.

TRIGLOCHIN L.
   T. maritima L. Arrow-grass. f. 186.
     Kings: Calverley. Long Island, Hulst. *Suffolk:* Miller & Young.

ALISMACEAE.

ALISMA L.
   A. Plantago-aquatica L. Water Plantain. f. 188.
     Frequent throughout the island.

SAGITTARIA L.
   S. latifolia Willd. f. 195.
     Long Island, Hulst. *Suffolk:* Miller & Young.
   S. rigida Pursh. f. 201.
     Queens: Richmond Hill.
   S. teres Wats. f. 202.
     Queens: Cold Spring, Hulst. *Suffolk:* Wading River, Miller.
   S. graminea Mx. f. 204.
     Long Island, Hulst. *Queens:* Jamaica, Ruger. *Suffolk:* Miller & Young.
   S. subulata (L) Buchenau. f. 206.
     Long Island, Hulst. *Queens:* Flushing, Leggett.

VALLISNERACIEAE

PHILOTRIA Raf.
   P. Canadensis (Mx.) Britton. f. 207.
     Long Island, Hulst.

VALLISNERIA L.
   V. spiralis L. Eel-grass. f. 208.
     Suffolk: Miller & Young.

GRAMINACEAE.

TRIPSACUM L.
  T. dactyloides L. Sesame Grass. f. 210.
  Kings: Forbell's Landing, Hulst.

ANDROPOGON L.
  A. scoparius Mx. f. 216.
  Queens: Rockaway Beach, Hulst. *Suffolk:* Miller & Young.
  A. scoparius maritimus (Chap.) Hack.
  Kings: Coney Island, Hulst.
  A furcatus Muhl. f. 219.
  Queens: Garden City, Hulst. *Suffolk:* Miller & Young.
  A. Virginicus L. f. 220.
  Queens: Maspeth, Hulst. *Suffolk:* Miller & Young.
  A. glomeratus (Walt.) B.S.P. f. 221.
  Kings: Forbell's Landing, Hulst. *Suffolk:* Miller & Young.

PASPALUM L.
  P. setaceum Mx. f. 231.
  Queens: Rockaway, Hulst. *Suffolk:* Manor, C. H. Peck; Miller & Young.

SYNTHERISMA Walt.
  S. sanguinalis (L.) Nash. f. 240.
  Kings: Brooklyn, Hulst. *Queens:* Rockaway Beach. *Suffolk:* Miller & Young.
  S. linearis (Krock) Nash. f. 241.
  Suffolk: Miller & Young.
  S. filiformis (L.) Nash. f. 242.
  Queens: Jamaica, Hulst. *Suffolk:* Miller & Young.

PANICUM L.
  P. Crus-galli L. f. 243.
  Common throughout the island.
  P. Walteri Pursh. f. 244.
  Common throughout the island.
  P. agrostidiforme Lam. f. 249.
  Queens: Maspeth, Hulst. *Suffolk:* Miller & Young.
  P. clandestinum L. f. 257.
  Kings: New Lots, Brainerd; Forbell's Landing, Cypress Hills, Hulst. *Suffolk:* Miller & Young.
  P. nitidum Lam. f. 263.
  Kings: Cypress Hill, Hulst.
  P. dichotomum L. f. 264.
  Queens: Rockaway Beach. *Suffolk:* Miller & Young.
  P. depauperatum Muhl. f. 268.
  Kings: Flatbush, Brainerd.

P. virgatum L. f. 270.
  Kings: Coney Island, Brainerd; Forbell's Landing, Hulst.
    Queens: Maspeth, Hulst. *Suffolk:* Miller & Young.
P. amarum L. f. 271.
  Suffolk: Miller & Young.
P. proliferum Lam. f. 273.
  Kings: Forbell's Landing, Hulst. *Suffolk:* Miller & Young.
P. capillare L. f. 274.
  Kings: Flatbush, Brainerd; Forbell's Landing, Hulst. *Queens:*
    Maspeth, Hulst. *Suffolk:* Miller & Young.
P. autumnale Bosc. f. 276.
  Kings: Cypress Hill, Hulst.
P. verrucosum Muhl. f. 278.
  Kings: Flatbush, Brainerd. *Suffolk:* Miller & Young.

IXOPHORUS Schlecht.
  I. glauca (L.) Nash. Bristly Fox-tail Grass. f. 281.
    Frequent throughout the island.
  I. viridis (L.) Nash. f. 282.
    Frequent throughout the island.

CENCHRUS L.
  C. tribuloides L. Bur-grass. f. 284.
    Frequent throughout the island.

ZIZANIA L.
  Z. aquatica L. Indian Rice. f. 286.
    Kings: Forbell's Landing, Hulst. *Suffolk:* Miller & Young.

HOMALOCENCHRUS Mieg.
  H. Virginica (Willd.) Britt. f. 287.
    Kings: Forbell's Landing, Hulst.
  H. oryzoides (L.) Oll. f. 288.
    Kings: Gowanus, Brainerd. *Queens:* Maspeth, Hulst. *Suffolk:*
      Miller & Young.

PHALARIS L.
  P. Canariensis L. Canary-grass. f. 292.
    Kings: Brooklyn, Hulst.

ANTHOXANTHUM L.
  A. odoratum L. Vernal Grass. f. 293.
    Common throughout the island.

SAVASTANA Schrank.
  S. odorata (L.) Scrib. Holy Grass. f. 294.
    Frequent throughout the island.

ARISTIDA L.
  A. dichotoma Mx. Poverty Grass. f. 297.
    Suffolk: Miller & Young.

A. gracilis Ell. f. 298.
    Suffolk: Miller & Young.
A. purpurascens Poir. f. 301.
    Suffolk: Miller & Young.
A. tuberculosa Nutt. f. 307.
    Suffolk: Miller & Young.

STIPA L.
S. avenacea L. Feather-grass. f. 311.
    Kings: New lots, Brainerd. *Suffolk:* Miller & Young.

MUHLENBERGIA Schreb.
M. Mexicana (L.) Trin. Drop-seed Grass. f. 321.
    Queens: Rockaway, Hulst. *Suffolk:* Miller & Young.
M. racemosa (Mx.) B.S.P. f. 322.
    Suffolk: Miller & Young.
M. diffusa Schreb. f. 327.
    Suffolk: Miller & Young.

BRACHYELYTRUM Beauv.
B. erectum (Schreb.) Beauv. f. 332.
    Kings: Flatbush, Brainerd.

HELEOCHLOA Host.
H. schoenoides (L.) Host. f. 333.
    Kings: Brooklyn, Hulst.

PHLEUM L.
P. pratense L. f. 335.
    Common throughout the island.

ALOPECURUS L.
A. geniculatus L. f. 337.
    Kings: Brooklyn, Hulst.
A. pratensis L. Meadow Foxtail. f. 338.
    Kings: Brooklyn, Calverley. ?

SPOROBOLUS R. Br.
S. asper (Mx.) Kunth. Dropseed grass. f. 341.
    Suffolk: Miller & Young.
S. vaginaeflorus (Torr.) Wood. f. 344.
    Suffolk: Miller & Young.
S. serotinus (Torr.), Gray. f. 355.
    Suffolk: Miller & Young.

CINNA L.
C. Arundinacea L. f. 360.
    Kings: Forbell's Landing, Hulst.
C. latifolia (Trev.) Griseb. f. 361.
    Kings: Forbell's Landing, Hulst. *Suffolk:* Miller & Young.

AGROSTIS L.
  A. alba L. White Bent-grass. f. 362.
    Kings: Forbell's Landing, Hulst. *Suffolk:* Miller & Young.
  A. alba vulgans (Wth.) Thunb.
    Suffolk: Miller & Young.
  A. perennans (Walt.) Tuck. Thin-grass. f. 367.
    Kings: Cypress Hill, Hulst.
  A. hyemalis (Walt.) B.S.P. Hair-grass. f. 368.
    Queens: Freshpond, Hulst.

CALAMAGROSTIS Adans.
  C. cinnoides (Muhl.) Scrib. f. 379.
    Queens: Maspeth, Hulst. *Suffolk:* Miller & Young.

AMMOPHILA Host
  A. arenaria (L.) Link. Sea Sand-reed. f. 380.
    Kings: Coney Island, Brainerd. *Queens:* Rockaway Beach, Hulst, J.

HOLCUS L.
  H. lanatus L. Velvet Grass. f. 384.
    Frequent throughout the island.

AIRA L.
  A. praecox L. f. 386.
    Kings: Brooklyn.

DESCHAMPSIA Beauv.
  D. flexuosa (L.) Trin. Hair-grass. f. 388.
    Queens: Jamaica, Brainerd. *Suffolk:* Greenport, Tillinghast, Miller & Young.

DANTHONIA DC.
  D. spicata (L.) Beauv. f. 397.
    Suffolk: Miller & Young.

CAPRIOLA Adans.
  C. Dactylon (L.) Kuntze. Bermuda grass. f. 400.
    Kings: Brooklyn, Brainerd, Hulst. *Queens:* Long Island City, Peck.

SPARTINA Schreb.
  S. cynosuroides (L.) Willd. Marsh Grass. f. 401.
    Kings: New Lots, Brainerd. *Suffolk:* Miller & Young.
  S. polystachya (Mx.) Ell. f. 402.
    Kings: Forbell's Landing, Hulst.
  S. patens (Ait.) Muhl. f. 403.
    Kings: New Lots, Brainerd, Forbell's Landing, Hulst. *Queens:* Rockaway Beach. *Suffolk:* Miller & Young.
  S. stricta (Ait.) Roth. f. 405.
    Kings: Coney Island, Brainerd; Forbell's Landing, Hulst.

S. stricta maritima (Lois.) Gray.
  Suffolk: Miller & Young.

ELEUSINE Gaertn.
  E. Indica (L.) Gaertn. Wire Grass. f. 415.
  Common throughout the island.

PHRAGMITES Trin.
  P. Phragmites (L.) Karst. f. 420.
  Kings: Forbell's Landing, Hulst. *Suffolk:* Miller & Young.

ERAGROSTIS Beauv.
  E. pilosa (L.) Beauv. f. 430.
  Kings: Brooklyn, Hulst.
  E. Purshii Schrad. f. 431.
  Kings: Brooklyn, Hulst.
  E. Eragrostis (L.) Karst. Eragrostis. f. 432.
  Kings: Brooklyn, Hulst. *Suffolk:* Miller & Young.
  E. major Host. f. 433.
  Kings: Brooklyn, Hulst.
  E. pectinacea spectabilis (Pursh) Gray. f. 436.
  Queens: Rockaway Beach. *Suffolk:* Miller & Young.

EATONIA Raf.
  E. Pennsylvanica (DC.) Gray. Eatonia. f. 442.
  Queens: Woodside, Hulst.

DISTICHLIS Raf.
  D. spicata (L.) Greene. Spike-grass. f. 454.
  Kings: Flatbush, Brainerd. *Suffolk:* Miller & Young.

DACTYLIS L.
  D. glomerata L. Orchard Grass. f. 457.
  Frequent throughout the island.

POA L.
  P. annua L. Low Spear-Grass. f. 459.
  Common throughout the island.
  P. compressa L. f. 461.
  Frequent throughout the island.
  P. pratensis L. Common meadow grass. f. 466.
  Common throughout the island.
  P. trivialis L. f. 468.
  Suffolk: Miller & Young.

PANICULARIA Fabr.
  P. Canadensis (Mx.) Kuntze. f. 485.
  Suffolk: Miller & Young.
  P. obtusa (Muhl.) Kuntze. f. 486.
  Suffolk: Miller & Young.

P. elongata (Torr.) Kuntze. Manna-grass. f. 487.
  Kings: Flatbush, Brainerd.
P. nervata (Willd.) Kuntze. f. 488.
  Kings: Flatbush, Brainerd. *Suffolk:* Miller & Young.
P. pallida (Torr.) Kuntze. f. 490.
  Suffolk: Miller & Young.
P. fluitans (L.) Kuntze. f. 491.
  Suffolk: Greenport, Tillinghast ; Miller & Young.

FESTUCA L.
  F. octoflora Walt. Fescue grass. f. 497.
    Queens: Rockaway Beach, Hulst. *Suffolk:* Greenport, Tillinghast ; Miller & Young.
  F. rubra L. f. 499.
    Suffolk: Miller & Young.
  F. ovina L. f. 500.
    Long Island, Brainerd. *Suffolk:* Miller & Young.
  F. ovina duriuscula (L.) Hack.
    Suffolk: Miller & Young.
  F. elatior L. f. 502.
    Queens: Winfield, Hulst.
  F. nutans Spreng. f. 504.
    Kings: New Lots, Brainerd.

BROMUS L.
  B. ciliatus L. f. 506.
    Queens: Rockaway, Hulst.
  B. tectorum L. f. 509.
    Suffolk: Miller & Young.
  B. secalinus L. Cheat. f. 514.
    Suffolk: Greenport, Tillinghast ; Miller & Young.
  B. racemosus L. f. 515.
    Kings: Flatbush, Brainerd.

AGROPYRON J. Gaertn.
  A. repens (L.) Beauv. f. 524.
    Suffolk: Miller & Young.

ELYMUS L.
  E. Virginicus L. Wild Rye. f. 534.
    Kings: Coney Island, Brainerd. *Queens:* Rockaway, Hulst.
    Suffolk: Miller & Young.
  E. Canadensis L. f. 535.
    Kings: Forbell's, Hulst. *Suffolk:* Miller & Young.

CYPERACEAE.

CYPERUS L.
  C. flavescens L. f. 543.
  Kings: Forbell's Landing, Hulst. *Queens:* Glen Cove.
  C. diandrus Torr. f. 544.
  Kings: New Lots, Brainerd ; Forbell's Landing, Hulst. *Queens:* Calverley, Glen Cove. *Suffolk:* Miller & Young.
  C. Nuttallii Eddy. f. 546.
  Kings: Forbell's Landing, Hulst. *Queens:* Hulst. *Suffolk:* Miller & Young.
  C. dentatus Torr. Galingale. f. 556.
  Kings: Forbell's, Hulst. *Queens:* Calverley. *Suffolk:* Miller & Young.
  C. speciosus Vahl. f. 562.
  Kings: Forbell's Landing, Hulst. *Queens:* Glen Cove. *Suffolk:* Miller & Young.
  C. strigosus L. f. 565.
  Frequent throughout the island.
  C. cylindricus (Ell.) Britt. f. 569.
  Queens: Rockaway Beach, Hulst, J.
  C. filiculmis Vahl. f. 571.
  Kings: Brooklyn, Hulst. *Queens:* Calverley ; Rockaway Beach. *Suffolk:* Miller & Young.
  C. Grayi Torr. f. 573.
  Suffolk: Miller & Young.

DULICHIUM L. C. Richard.
  D. arundinaceum (L.) Britt f. 576.
  Kings: Calverley. *Queens:* Newtown, Hulst ; Glen Cove. *Suffolk:* Miller & Young.

ELEOCHARIS R. Br.
  E. Robbinsii Oakes. f. 579.
  Suffolk: Miller & Young.
  E. olivacea Torr. f. 581.
  Suffolk: Miller & Young.
  E. ovata (Roth.) Roem. & Schult. f. 584.
  Kings: New Lots, Brainerd, Cypress Hills, Hulst. *Suffolk:* Miller & Young.
  E. palustris (L.) Roem. & Schult. f. 586.
  Queens: Woodhaven, Hulst ; Glen Cove. *Suffolk:* Miller & Young.
  E. acicularis (L.) Roem. & Schult. f. 587.
  Kings: New Lots, Brainerd ; Cypress Hill, Hulst. *Queens:* Cold Spring Harbor, Hulst. *Suffolk:* Miller & Young.

E. melanocarpa Torr. f. 592.
Suffolk: Miller & Young.
E. tenuis (Willd.) Schult. f. 595.
Kings: Carnarsie, Brainerd. *Queens:* Woodside, Hulst.
E. intermedia (Muhl.) Schult. f. 597.
Kings: Forbell's Landing, Hulst. *Queens:* Calverley.
E. rostellata Torr. f. 598.
Kings: Calverley. *Suffolk:* Miller & Young.

PSILOCARYA Torr.
P. nitens (Vahl) Wood. f. 601.
Suffolk: Wading River, Miller.

STENOPHYLLUS Raf.
S. capillaris (L.) Britt. f. 603.
Kings: Cypress Hill, Hulst. *Suffolk:* Miller & Young.

FIMBRISTYLIS Vahl.
F. castanea (Mx.) Vahl. f. 605.
Suffolk: Miller & Young.
F. autumnalis (L.) Roem. & Schult. Fimbristylis. f. 608.
Kings: Cypress Hills, Hulst. *Queens:* Glen Cove. *Suffolk:* Miller & Young.

SCIRPUS L.
S. nanus Spreng. f. 609.
Suffolk: Miller & Young.
S. caespitosus L. f. 611.
Kings: Forbell's Landing, Hulst. *Queens:* Calverley.
S. planifolius Muhl. f. 613.
Kings: Flatbush, Brainerd.
S. subterminalis Torr. f. 614.
Queens: Cold Spring Harbor, Hulst. *Suffolk:* Miller & Young.
S. debilis Pursh. f. 616.
Suffolk: Miller & Young.
S. Americanus Pers. f. 618.
Kings: Canarsie, Brainerd. *Queens:* Maspeth, Hulst. *Suffolk:* Miller & Young.
S. Olneyi Gray. f. 619.
Suffolk: Miller & Young.
S. Torreyi Olney. f. 620.
Kings: Forbell's Landing, Hulst. *Queens:* Rockaway Beach, Hulst, J.
S. lacustris L. f. 623.
Queens: Maspeth, Hulst; Glen Cove. *Suffolk:* Miller & Young.
S. atrovirens Muhl. Bulrush. f. 630.
Kings: Forbell's Landing, Hulst; Canarsie, Brainerd. *Queens:* Maspeth, Hulst; Glen Cove. *Suffolk:* Miller & Young.

S. polyphyllus Vahl. f. 632.
Queens: Cold Spring Harbor, Hulst. Suffolk: Miller & Young.
S. cyperinus Eriophorum (Michx.) Britt. f. 636.
Kings: Cypress Hills, Hulst; Flatlands, Brainerd. Suffolk: Miller & Young.

ERIOPHORUM L.
E. polystachyon L. Cotton Grass. f. 641.
Queens: Calverley. Suffolk: Miller & Young.
E. gracile Koch. f. 642.
Suffolk: Miller & Young.
E. Virginicum L. f. 643.
Frequent throughout the island.

FUIRENA Rottb.
F. squarrosa Mx. Umbrella Grass. f. 644.
Suffolk: Miller & Young.

HEMICARPHA Nees & Am.
H. micrantha (Vahl) Britt. f. 646.
Long Island, Calverley.

RYNCHOSPORA Vahl.
R. corniculata macrostachya (Torr.) Britt. f. 648.
Suffolk: Miller & Young.
R. alba (L.) Vahl. Beak-rush. f. 651.
Kings: Forbell's Landing, Hulst. Suffolk: Miller & Young.
R. glomerata (L.) Vahl. f. 654.
Kings: Coney Island, Brainerd. Suffolk: Miller & Young.
R. axillaris (Lam.) Britt. f. 655.
Suffolk: Miller & Young.
R. fusca (L.) Roem. & Sch. f. 656.
Suffolk: Miller & Young.

CLADIUM P. Br.
C. mariscoides (Muhl.) Torr. Twig-rush. f. 661.
Kings: Coney Island, Brainerd; Rockaway Beach. Suffolk: Miller & Young.

SCLERIA Berg.
S. reticularis Mx. Nut-rush. f. 664.
Suffolk: Miller & Young.

CAREX L.
C. folliculata L. f. 674.
Common throughout the island.
C. intumescens Rudge. f. 675.
Queens: Valley Stream. Suffolk: Miller & Young.
C. lurida Wahl. f. 693.
Frequent throughout the island.

C. Pseudo-Cyperus L. f. 697.
Queens: Valley Stream.
C. comosa Boott. f. 698.
Suffolk: Miller & Young.
C. squarrosa L. f. 700.
Kings: Forbell's Landing, Hulst.
C. scabrata Schw. f. 707.
Queens: Richmond Hill.
C. vestita Willd. f. 708.
Kings: New Lots, Brainerd. *Suffolk:* Miller & Young.
C. Walteriana Bailey. f. 709.
Suffolk: Miller & Young.
C. filiformis L. f. 712.
Kings: New Lots, Brainerd.
Var. lanuginosa (Mx.).
Suffolk: Miller & Young.
C. fusca All. f. 718.
Kings: Forbell's Landing, Hulst.
C. stricta Lam. f. 719.
Kings: Canarsie, Brainerd. *Queens:* Maspeth, Hulst. *Suffolk:* Miller & Young.
C. Goodenovii J. Gay. f. 724.
Kings: New Lots, Brainerd.
C. crinita Lam. f. 739.
Kings: New Lots, Brainerd. *Queens:* Cold Spring Harbor, Hulst; Valley Stream. *Suffolk:* Miller & Young.
C. virescens Muhl. f. 743.
Kings: Prospect Park. *Suffolk:* Miller & Young.
C. longirostris Torr. f. 752.
Kings: Coney Island, Brainerd.
C. glaucodea Tuck. f. 762.
Queens: Richmond Hill, Hulst.
C. granularis Muhl f. 763.
Long Island, C. H. Hall.
C. extensa Good. f. 765.
Kings: Coney Island, Calverley, notes, Britton & Brown.
C. pallescens L. f. 768.
Norwich, C. H. Hall.
C. laxiflora Lam. f. 777.
Frequent throughout the island.
C. l. blanda (Dewey) Bost.
Kings: Prospect Park.
C. l. patulifolia (Dewey) Carey.
Suffolk: Miller & Young.

C. digitalis Willd. f. 779.
Long Island, C. H. Hall.
C. laxiculmis Schwein. f. 783.
Kings: New Lots, Brainerd; Prospect Park.
C. Pennsylvanica Lam. f. 795.
Frequent throughout the island.
C. varia Muhl. f. 796.
Kings: Cypress Hills, Hulst. *Queens*: Jamaica, Brainerd.
C. nigro-marginata Schwein. f. 800.
Suffolk: Miller & Young.
C. umbellata Schk. f. 801.
Kings: New Lots, Brainerd. *Suffolk*: Miller & Young.
C. scirpoidea Mx. f. 805.
Queens: Valley Stream.
C. exilis Dewey. f. 816.
Suffolk: Miller & Young.
C. stipata Muhl. f. 823
Kings: Coney Island, Brainerd; Forbell's Landing, Hulst. *Suffolk*: Miller & Young.
C. alopecoidea Tuck. Sedge. f. 828.
Queens: Valley Stream.
C. vulpinoidea Mx. f. 830.
Queens: Jamaica, Brainerd. *Suffolk*: Miller & Young.
C. rosea Schk. f. 835.
Kings: Flatlands, Brainerd; Prospect Park. *Suffolk*: Miller & Young.
C. r. radiata Dewey.
Kings: Prospect Park.
C. sparganoides Muhl. f. 839.
Kings: Coney Island, Brainerd; Forbell's Landing, Hulst.
Queens: Valley Stream.
C. cephalophora Muhl. f. 841.
Kings: Forbell's Landing, Hulst. *Suffolk*: Miller & Young.
C. Muhlenbergii Schk. f. 843.
Suffolk: Miller & Young.
C. sterilis Willd. f. 844.
Suffolk: Miller & Young.
C. canescens L. f. 847.
Frequent throughout the island.
C. trisperma Dewey. f. 855.
Kings: Prospect Park.
C. scoparia Schk. f. 863.
Frequent throughout the island.
C. straminea Willd. f. 868
Kings: New Lots, Brainerd. *Suffolk*: Miller & Young.

C. silicea Olney. f. 869.
 Suffolk: Miller & Young.
C. alata Torr. f. 872.
 Kings: New Lots, Brainerd. *Suffolk:* Miller & Young.
C. albolutescens Schwein. f. 873.
 Kings: Flatlands, Brainerd.

ARACEAE.

ARISAEMA Mart.
 A. triphyllum (L.) Torr. Jack-in-the-Pulpit. f. 876.
 Common throughout the island.

PELTANDRA Raf.
 P. Virginica (L.) Kunth. Arrow Arum. f. 878
 Long Island, Hulst. *Queens:* Calverley, Flushing, Newtown. Bisky. *Suffolk:* Miller & Young.

CALLA L.
 C. palustris L. Water Arum. f. 880.
 Long Island, Hulst.

SPATHYEMA Raf.
 S. foetida (L.) Raf. Skunk Cabbage. f. 881.
 Common throughout.

ORONTIUM L.
 O. aquaticum L. Golden Club. f. 882.
 Queens: Jamaica, Bisky. *Suffolk:* Miller & Young.

ACORUS L.
 A. Calamus L. Calamus. f. 883.
 Frequent throughout the island.

LEMNACEAE.

SPIRODELA Schleid.
 S. polyrhiza (L.) Schleid. Duck-weed. f. 884.
 Kings: Cypress Hills, Hulst. *Queens:* Richmond Hill, Hulst.

LEMNA L.
 L. trisulca L. f. 885.
 Queens: Newtown, M. Ruger.
 L. Valdiviani Phillippi. f. 886.
 Suffolk: Miller & Young.
 L. minor L. Duck-weed. f. 888.
 Common throughout.

XYRIDACEAE.

XYRIS L.
  X. flexuosa Muhl. Yellow-eyed grass. f. 893.
    Kings: Forbell's Landing, Hulst. *Queens:* Hempstead, Jamacia, Oyster Bay, Bisky. *Suffolk:* Miller & Young.
  X. Caroliniana Walt. f. 896.
    Frequent throughout the island.

ERIOCAULONACEAE.

ERIOCAULON L.
  E. septangulare With. f. 899.
    Queens: Calverley; N. Hempstead, Bisky. *Suffolk:* Miller & Young.
  E. decangulare L. Pipewort. f. 901.
    Queens: Calverley; Ridgewood, Bisky.

COMMELINACEAE.

COMMELINA L.
  C. Virginica L. Day flower. f. 909.
    Kings: Cypress Hill, Hulst; Prospect Park. *Queens:* Flushing, Bisky.

TRADESCANTIA L.
  T. Virginiana L. Spiderwort. f. 910.
    Kings: Cypress Hills, Hulst; Calverley, Brooklyn.

PONTEDERIACEAE.

PONTEDERIA L.
  P. cordata L. Pickerel Weed. f. 915.
    Frequent in water throughout the island.
  P. c. lancifolia (Muhl.) Morong.
    Queens: Calverley.

JUNCACEAE.

JUNCUS L.
  J. effusus L. f. 919.
    Frequent throughout the island.
  J. Roemerianus Scheele. f. 924.
    Long Island, Hulst. *Kings:* Coney Island, Vid. N. L. Britton, Plants of New Jersey, p. 249.
  J. maritimus Lam. f. 925.
    Kings: Coney Island, Brainerd; Britton & Brown, p. 384.
  J. bufonius L. f. 926.
    Queens: Cold Spring Harbor, Hulst. *Suffolk:* Miller & Young.

J. trifidus L. f. 927.
: Queens: Sand's Point, Bisky. (?)
J. Gerardi Lois. f. 928.
: *Kings*: Calverley; Coney Island, Brainerd. *Queens*: Woodhaven, Leggett. *Suffolk*: Greenport, Tillinghast; Miller & Young.
J. tenuis Willd. f. 929.
: Frequent throughout the island.
J. Greenei Oakes & Tuck. f. 932.
: *Queens*: Hicksville, Hulst. *Suffolk*: Miller & Young.
J. marginatus Rostk. f. 935.
: *Suffolk*: Miller & Young.
J. pelocarpus E. Meyer. f. 942.
: *Suffolk*: Miller & Young.
J. militaris Big. f. 944.
: *Suffolk*: Miller & Young.
J. nodosus L. f. 947.
: *Kings*: Flatlands, Brainerd. *Queens*: Plandome, Eddy.
J. scirpoides Lam. f. 952.
: *Suffolk*: Yaphank, Peck, N. Y. St. Rept. 1890.
J. Canadensis J. Gay. f. 955.
: *Kings* · Canarsie, Brainerd. *Suffolk*: Miller & Young.
J. acuminatus Mx. Rush. f. 956.
: *Kings*: Hulst. *Queens*: Astoria, Jamaica, Leggett. *Suffolk*: Miller & Young.
J. a. debilis (Gray) Eng.
: *Kings*: Calverley.

JUNCOIDES Adans.
J. pilosum (L.) Kuntze. f. 959.
: Frequent throughout the island.
J. campestre (L.) Kuntze. f. 965.
: Frequent throughout the island.

MELANTHACEAE.

CHAMAELIRIUM Willd.
C. luteum (L.) A. Gray. Blazing Star. f. 972.
: *Queens*: Richmond Hill, Rusby; Flushing, Jamaica, Bisky; Valley Stream.

CHROSPERMA Raf.
C. muscaetoxicum (Walt.) Kuntze. Fly Poison. f. 973.
: *Queens*: Hempstead, Bisky; Valley Stream, Hulst, J.

MELANTHIUM L.
M. Virginicum L. Bunch flower. f. 981.
: *Queens*: Richmond Hill, *fide* J. H. Rudkin.

VERATRUM L.
   V. viride Ait. White Hellebore. f. 985.
   Frequent throughout the island.

UVULARIA L.
   U. perfoliata L. Bellwort. f. 986.
   Frequent throughout the island.
   U. sessilifolia (L). f. 988.
      Kings: Hulst; Prospect Park. *Queens:* Flushing, Bisky. *Suffolk:* Miller & Young.

LILIACEAE.

HEMEROCALLIS L.
   H. fulva L. Day Lily. f. 990.
   Frequently escaped from gardens throughout the island.

ALLIUM L.
   A. tricoccum Ait. f. 992.
      Queens: Flushing, Bisky.
   A. vineale L. Wild Garlic. f. 996.
   Frequent throughout the island.
   A. Canadense Kalm. Meadow Garlic. f. 997.
   Frequent throughout the island.

LILIUM L.
   L. Philadelphicum L. Wild Red Lily. f. 1003.
   Frequent throughout the island.
   L. Canadense L. Wild Yellow Lily. f. 1006.
   Frequent throughout the island.
   L. surperbum L. Turk's Cap Lily. f. 1008.
      Kings: Forbell's Landing, Hulst. *Queens:* Calverley, Flushing, Newtown, Oyster Bay, Bisky. *Suffolk:* Greenport, Tillinghast; E. Hampton, Mrs. L. D. Pychowska; Miller & Young.

ERYTHRONIUM L.
   E. Americanum Ker. Yellow Adders' Tongue. f. 1012.
      Kings: Prospect Park. *Queens:* Calverley, Maspeth, Hulst; Flushing, Hempstead, Newtown, Bisky; Richmond Hill.

ORNITHOGALUM L.
   O. umbellatum L. Star of Bethlehem. f. 1019.
   Common throughout the island.

MUSCARI L.
   M. botryoides (L.) Mill. Grape Hyacinth. f. 1021.
      Kings: Cypress Hills, Hulst; Prospect Park. *Queens:* Flushing, N. Hempstead, Bisky. *Suffolk:* Miller & Young.

ALETRIS L.
   A. farinosa L. Star-grass. f. 1023.
     Kings: Forbell's Landing, Hulst. *Queens:* Calverley, Hicksville, Hulst; Hempstead, Jamaica, Bisky. *Suffolk:* Miller & Young; Sag Harbor, E. Hampton, Mrs. L. D. Pychowska.

CONVALLARIACEAE.

ASPARAGUS L.
   A. officinalis L. Asparagus. f. 1028.
     Common throughout the island.

VAGNERA Adans.
   V. racemosa (L.) Morong. False Solomon's Seal. f. 1031.
     Common throughout the island.
   V. stellata (L.) Morong. f. 1032.
     Frequent throughout the island.

UNIFOLIUM Adans.
   U. Canadense (Desf.) Greene. False Lily-of-the-Valley. f. 1034.
     Common throughout the island.

POLYGONATUM Adans.
   P. biflorum (Walt.) Ell. Solomon's Seal. f. 1039.
     Frequent throughout the island.
   P. commutatum (R. & S.) Dietr. f. 1040.
     Queens: Richmond Hill, Hulst; Flushing, Bisky. *Suffolk:* Three Mile Harbor, Mrs. L. D. Pychowska; Miller & Young.

CONVALLARIA L.
   C. majalis L. Lily-of-the-Valley. f. 1041.
     Kings: Prospect Park; Cypress Hills, Hulst. *Queens:* Flushing, Bisky. (Escaped.)

MEDEOLA L.
   M. Virginica L. Indian Cucumber Root. f. 1042.
     Infrequent throughout the island.

TRILLIUM L.
   T. erectum L. f. 1047.
     Queens: Glen Cove, Lawrenceville, Ruger; N. Hempstead, Bisky; Coles.
   T. cernuum L. Wake Robin. f. 1048.
     Queens: Jamaica, Hulst.
   T. undulatum Willd. f. 1049.
     Queens: Glen Cove, Coles.

SMILACEAE.

SMILAX L.
   S. herbacea L. Carrion Flower. f. 1050.
     Frequent throughout the island.

Var. pulverulenta (Mx.) Gray.
  Queens: Greenpoint, Allen; Leggett.
S. glauca Walt. f. 1053.
  Frequent throughout the island.
S. rotundifolia L. Common Greenbrier. f. 1054.
  Frequent throughout the island.

HAEMODORACEAE.

GYROTHECA Salisb.
  G. capitata (Walt.) Morong. f. 1061.
  Suffolk: Manor, C. H. Peck, N. Y. St. Repts. 1890.

AMARYLLIDACEAE.

HYPOXIS L.
  H. hirsuta (L.) Coville. Yellow-eyed grass. f 1066.
  Common throughout the island.

DIOSCOREACEAE.

DIOSCOREA L.
  D. villosa L. Yam. Root. f. 1068.
    Kings: Forbell's Landing, Hulst. *Queens:* Flushing, Bisky; Calverley. *Suffolk:* Southhold, Tillinghast; Miller & Young.

IRIDACEAE.

IRIS L.
  I. versicolor L. Blue flag. f. 1069.
    Frequent throughout the island.
  I. prismatica Pursh. f. 1074.
    Frequent throughout the island.
GEMMINGIA Fabr.
  G. Chinensis (L.) Kuntze. Blackberry Lily. f. 1082.
    Queens: Richmond Hill, Hulst; Jamaica, Middleville, Ruger.
SISYRINCHIUM L.
  S. angustifolium Mull. Blue-eyed grass. f. 1085.
    Common throughout the island.

ORCHIDACEAE.

CYPRIPEDIUM L.
  C. acaule Ait. Moccasin flower. f. 1089.
    Kings: Forbell's, Hulst. *Queens:* Cold Spring, Hulst, Calverley, Flushing, Hempstead, Jamaica, Oyster Bay, Bisky. *Suffolk:* Greenport, Tillinghast, Miller & Young.

ORCHIS L.
   O. spectabilis L. Orchis. f. 1094.
     Long Island, Hulst.

HABENARIA Willd.
   H. clavellata (Mx.) Spreng. f. 1104.
     Frequent throughout the island.
   H. flava (L.) Gray. f. 1105.
     Queens: Calverley ? Glen Cove, Coles.
   H. cristata (Mx.) R. Br. f. 1106.
     Kings: New Lots, Brainerd. *Queens:* Calverley.
   H. ciliaris (L.) R. Br. f. 1107.
     Kings: New Lots, Brainerd, Forbell's, Hulst. *Queens:* Calverley, Hicksville, Hulst; Glen Cove, Coles, J. *Suffolk:* Springs, Mrs. L. D. Pychowska.
   H. blephariglottis (Willd.) Torr. f. 1108.
     Kings: Forbell's, Hulst. *Queens:* Calverley, Jamaica, Bisky. *Suffolk:* Springs, Mrs. L. D. Pychowska, Miller & Young.
   H. lacera (Mx.) R. Br. f. 1109.
     Kings: New Lots, Brainerd, Hulst; Calverley. *Queens:* Flushing, Oyster Bay, Bisky. *Suffolk:* Greenport, Tillinghast, Miller & Young.
   H. psycodes (L.) Gray. f. 1112.
     Queens: Jamaica, Brainerd, Flushing, Oyster Bay, Bisky.

POGONIA Juss.
   P. ophioglossoides (L.) Ker. Pogonia. f. 1114.
     Frequent throughout the island.
   P. verticellata (Willd.) Nutt. f. 1117.
     Kings: Forbell's Landing, Hulst. *Queens:* Calverley, Aqueduct, Hulst; Bowery Bay, Ruger; Jamaica, Leggett. *Suffolk:* Miller & Young.

ARETHUSA L.
   A. bulbosa L. Arethusa. f. 1119.
     Kings: New Lots, Brainerd; Near Canarsie, Dr. Eccles.? *Queens:* Calverley, Hempstead; Oyster Bay, Bisky; Valley Stream, J. *Suffolk:* Miller & Young.

GYROSTACHYS Pers.
   G. plantaginea (Raf.) Britton. f. 1122.
     Queens: Glen Cove, Bisky. ?
   G. cernua (L.) Kuntze. Ladies' Tresses. f. 1123.
     Frequent throughout the island.
   G. praecox (Walt.) Kuntze. f. 1125.
     Frequent thoughout the island.

G. simplex (A. Gray) Kuntze. f. 1126.
Kings: Cedarhurst, Torrey Club, 1891. Queens: Hempstead, Bisky.
G. gracilis (Bigel.) Kuntze. f. 1127.
Frequent throughout the island.

PERAMIUM Salisb.
P. repens (L.) Salisb. f. 1131.
Queens: Calverley, Jamaica, Brainerd.
P. pubescens (Willd.) MacM. Rattlesnake Plantain. f. 1132.
Frequent throughout the island.

ACHROANTHES Raf.
A. unifolia (Mx.) Raf. Adder's mouth. f. 1135.
Queens: Cold Spring, Hulst; Jamaica, Ruger. Suffolk: Miller & Young.

LEPTORCHIS Du Petit Thouars.
L. liliifolia (L.) Kuntze. Tway-blade. f. 1136.
Kings: Calverley, Cypress Hills, Hulst. Queens: Calverley, Jamaica, N. Hempstead; Oyster Bay, Bisky. Suffolk: Miller & Young.
L. Loeselii (L.) MacM. f. 1137.
Queens: Flushing, Newtown, Oyster Bay, Bisky.

CORALLORHIZA R. Br.
C. odontorhiza (Willd.) Nutt. Coral root. f. 1140.
Kings: Cypress Hills, Hulst. Queens: College Point, Schrenk. Suffolk: Miller & Young.
C. multiflora Nutt. f. 1142.
Kings: Calverley. Queens: Flushing, N. Hempstead; Oyster Bay, Bisky. Suffolk: Miller & Young.

LIMODORUM L.
L. tuberosum L. Calopogon. f. 1145.
Frequent throughout the island.

DICOTYLEDONES.

SAURURACEAE.

SAURURUS L.
S. cernuus L. Lizards-Tail. f. 1148.
Queens: Glen Cove, Coles.

JUGLANDACEAE.

JUGLANS L.
J. nigra L. Black Walnut. f. 1149.
Kings: Brooklyn; Long Island, Hulst. Queens: Richmond Hill; Throughout, Bisky. Suffolk: Miller & Young.

J. cinerea L. Butternut. f. 1150.
  Queens: Evergreens, Hulst ; Richmond Hill ; Flushing, Oyster Bay, Bisky. *Suffolk:* Miller & Young.

HICOREA Raf
  H. minima (Marsh) Britt. f. 1152.
  Queens: Flushing, Bisky.
  H. ovata (Mill.) Britt. Shell-bark. f. 1154.
  Suffolk: Greenport, Tillinghast ; Miller & Young.
  H. alba (L.) Britt. f. 1156.
  Long Island, Hulst. *Queens:* Throughout, Bisky. *Suffolk:* Greenport, Tillinghast ; Miller & Young.
  H. glabra (Mill.) Britt. Pig-nut. f. 1158.
  Kings: Calverley. Long Island, Hulst. *Queens:* Flushing, Bisky. *Suffolk:* Miller & Young.

MYRICACEAE.

MYRICA L.
  M. Gale L. Sweet Gale. f. 1159.
  Queens: Jamaica, Rudkin. *Suffolk:* Miller & Young.
  M. cerifera L. f. 1160.
  Frequent throughout the island.

COMPTONIA Banks.
  C. peregrina (L.) Coulter. Sweet-fern. f. 1162.
  Frequent throughout the island.

SALICACEAE.

POPULUS L.
  P. alba L. f. 1164.
  Kings: Calverley ; Long Island, Hulst. *Queens:* Flushing, Newtown, Bisky.
  P. balsamifera candicans (Ait.) Gray. f. 1165.
  Kings: Brooklyn, Hulst. *Queens:* College Point, Schrenk.
  P. heterophylla L. f. 1168.
  Frequent throughout the island.
  P. grandidentata Mx. f. 1169.
  Queens: Newtown, Bisky. *Suffolk:* Greenport, Tillinghast ; Miller & Young.
  P. tremuloides Mx f 1170.
  Frequent throughout the island.
  P. dilatata L.
  Frequent throughout the island.
  P. deltoides Marsh. f. 1172.
  Kings: Brooklyn, Calverley.

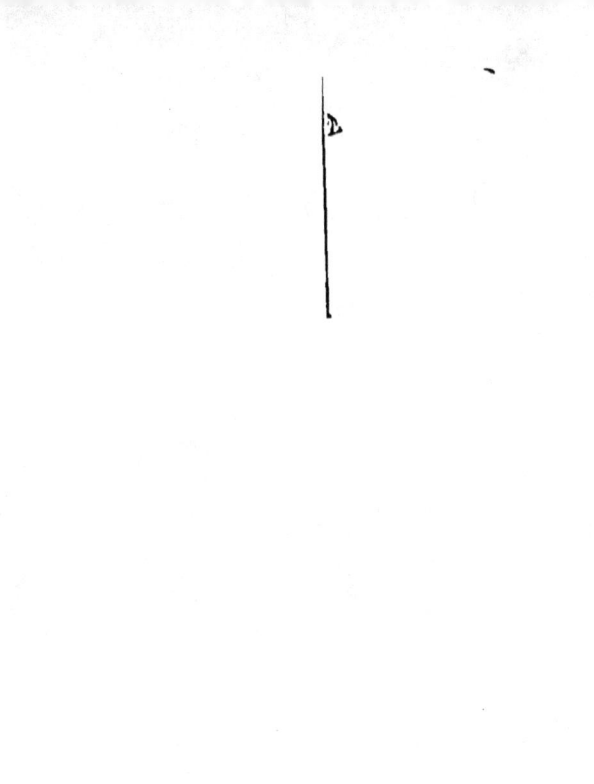

SALIX L.
  S. nigra Marsh. f. 1173.
    Long Island, Hulst. *Queens:* Woodside, E. N. Day.
  S. fragilis L. f. 1177.
    Long Island, Hulst.
  S. alba vitellina (L.) Koch. f. 1178.
    Queens: Glen Cove. *Suffolk:* Miller & Young.
  S. Babylonica L. f. 1179.
    Frequent throughout the island.
  S. humilis Muhl. f. 1185.
    Kings: Calverley. Long Island, Hulst. *Queens:* Long Island City, T. F. Allen. *Suffolk:* Miller & Young.
  S. tristis Ait. f. 1186.
    Long Island, Hulst. *Queens:* Calverley. *Suffolk:* Miller & Young.
  S. discolor Muhl. f. 1187.
    Long Island, Hulst. *Kings:* New Lots, Morong. *Queens:* E. G. Britton.
  S. sericea Marsh. f. 1188.
    Long Island, Hulst.
  S. cordata Muhl. f. 1198.
    Kings: Prospect Park.

BETULACEAE.

CARPINUS L.
  C. Caroliniana Walt. f. 1207.
    Queens: Flushing, N. Hempstead; Oyster Bay, Bisky.

OSTRYA Scop.
  O. Virginiana (Mill.) Willd. f. 1208.
    Kings: Prospect Park. *Queens:* Flushing, N. Hempstead; Oyster Bay, Bisky. *Suffolk:* Miller & Young.

CORYLUS L.
  C. Americana Walt. f. 1209.
    Kings: Cypress Hills, Hulst. *Queens:* Richmond Hill; Flushing, Newtown, N. Hempstead, Oyster Bay, Bisky. *Suffolk:* Miller & Young.

BETULA L.
  B. populifolia Marsh. f. 1211.
    Frequent throughout the island.
  B. nigra L. f. 1214.
    Long Island, Hulst. *Queens:* Oyster Bay, Bisky; Glen Cove. *Suffolk:* Miller & Young.
  B. lenta L. f. 1215.
    Frequent throughout the island.

ALNUS Gaertn.
  A. incana (L.) Willd. f. 1221.
    Frequent throughout the island.
  A. rugosa (Du Roi) Koch. f. 1222.
    Frequent throughout the island.
  A. glutinosa (L.) Medic. f. 1223.
    Kings: Prospect Park. *Queens:* Flushing, Bisky.

FAGACEAE.

FAGUS L.
  F. Americana Sweet. f. 1225.
    Kings: Prospect Park.
  F. atrapunicea (Marsh) Sudw.
    Frequent throughout the island.

CASTANEA Adans.
  C. dentata (Marsh) Borkh. f. 1226.
    Frequent throughout the island.

QUERCUS L.
  Q. rubra L. f. 1228.
    Queens: Flushing, Bisky. *Suffolk:* Greenport, Tillinghast; Miller & Young.
  Q. palustris Du Roi. f. 1229.
    Kings: Prospect Park. *Queens:* Calverley, Flushing, Newtown, N. Hempstead, Oyster Bay, Bisky. *Suffolk:* Greenport, Tillinghast, Miller & Young.
  Q. coccinea Wang. f. 1231.
    Long Island, Hulst; Prospect Park. *Queens:* Flushing, Oyster Bay, Bisky. *Suffolk:* Greenport, Tillinghast; Miller & Young.
  Q. velutina Lam. f. 1232.
    Queens: Flushing, Jamaica, Oyster Bay, Bisky. *Suffolk:* Greenport, Tillinghast; Miller & Young.
  Q. nana (Marsh) Sarg. f. 1234.
    Long Island, Hulst. *Queens:* Jamaica, Oyster Bay, Bisky. Jamaica, Lyosset, Schrenk. *Suffolk:* East Hampton, Mrs. L. D. Pychowska; Miller & Young.
  Q. nigra L. f. 1236.
    Long Island, Hulst. *Queens:* Oyster Bay, Bisky. *Suffolk:* Miller & Young.
  Q. Phellos L. f. 1237.
    Suffolk: Miller & Young.
  Q. alba L. f. 1240.
    Long Island, Hulst. *Queens:* Cedarhurst, Torrey Club; Bisky. *Suffolk:* Greenport, Tillinghast; Miller & Young.

Q. minor (Marsh) Sarg. f. 1241.
 Queens: Hempstead; Oyster Bay, Bisky; *Suffolk:* Greenport, Tillinghast.
Q. platanoides (Lam.) Sudw. f. 1244.
 Queens: Flushing, Bisky. *Suffolk:* Greenport, Tillinghast.
Q. prinus L. f. 1246.
 Long Island, Hulst. *Queens:* Hempstead, Oyster Bay, Bisky. *Suffolk:* Wading River, Hulst; Greenpoint, Tillinghast; Miller & Young.
Q. acuminata (Mx.) Sarg. f. 1247.
 Queens: Bisky, no locality; Locust Valley, Glen Cove, Oyster Bay.
Q. Muhlenbergii humilis (Marsh) Butt.
 Kings: Forbell's Landing, Hulst; Ridgewood, Brainerd. *Queens:* Hempstead, Bisky. *Suffolk:* East Hampton, Mrs. L. D. Pychowska; Miller & Young.

ULMACEAE.

ULMUS L.
 U. Americana L. f. 1250.
  Frequent throughout the island.
 U. pubescens Walt. f. 1253.
  Long Island, Hulst. *Kings:* Prospect Park. *Queens:* Glen Cove, Coles; Oyster Bay, Bisky.

CELTIS L.
 C. occidentalis L. f. 1255.
  Kings: Brooklyn, Forbell's Landing, Hulst. *Queens:* Flushing, Newtown, Oyster Bay, Bisky; Glen Cove.

MORACEAE.

MORUS L.
 M. rubra L. f. 1257.
  Long Island, Hulst. *Kings:* Calverley. *Queens:* Flushing, Newtown, Bisky. *Suffolk:* Miller & Young.
 M. alba L. f. 1258.
  Long Island, Hulst. *Kings:* Brooklyn. *Queens:* Newtown, Bisky; Astoria, Rudkin, Glen Cove.

TOXYLON Raf.
 T. pomiferum Raf. f. 1259.
  Cultivated.

BROUSSONETIA L'Her.
 B. papyrifera (L.) Vent.
  Queens: Escaped from cultivation.

HUMULUS L.
  H. Lupulus L. f. 1261.
    Frequent throughout the island.

CANNABIS L.
  C. sativa L. f. 1262.
    Kings: Calverley. *Queens:* Newtown, Hulst; Jamaica, Newtown, Bisky.

URTICACEAE.

URTICA L.
  U. dioica L. f. 1263.
    Queens: Woodside, Hulst; Richmond Hill, Flushing, Newtown, Oyster Bay, Bisky.
  U. gracilis Ait. f. 1264.
    Long Island, Hulst. *Queens:* Flushing, N. Hempstead, Oyster Bay, Bisky. *Suffolk:* Miller & Young.

URTICASTRUM Fabric.
  U. divaricatum (L.) Kuntze. f. 1267.
    Frequent throughout the island.

ADICEA Raf.
  A. pumila (L.) Raf. f. 1268.
    Frequent throughout the island.

BOEHMERIA Jacq.
  B. cylindrica (L.) Willd. f. 1269.
    Frequent throughout the island.

SANTALACEAE.

COMANDRA Nutt.
  C. umbellata (L.) Nutt. f. 1273.
    Frequent throughout the island.

ARISTOLOCHIACEAE.

ASARUM L.
  A. Canadense L. f. 1277.
    Queens: Newtown, Bisky.

ARISTOLOCHIA L.
  A. Serpentaria L. f. 1281.
    Queens: Jamaica, Brainerd, Merriam; Glen Cove, Coles.
  A. Clematitis L. f. 1282.
    Kings: Flushing, Britton & Brown.

POLYGONACEAE.

RUMEX L.
  R. Acetosella L.   f. 1299.
   Common throughout the island.
  R. hastatulus Muhl.   f. 1300.
   Suffolk: Miller & Young.
  R. verticillatus L.   f. 1304.
   Long Island, Hulst.
  R. Britannica L.   f. 1307.
   Suffolk: Miller & Young.
  R. crispus L.   f. 1309.
   Frequent throughout the island.
  R. conglomeratus Murray.   f. 1310.
   Queens: Flushing, Bisky.
  R. obtusifolius L.   f. 1313.
   Frequent throughout the island.

FAGOPYRUM Gaertn.
  F. Fagopyrum (L.) Karst.   f. 1316.
   Frequent throughout the island.

POLYGONUM L.
  P. amphibium L.   f. 1319.
   Kings: Forbell's Landing, Hulst.   *Queens:* Flushing, Newtown, Oyster Bay, Bisky.
  P. incarnatum Ell.   f. 1323.
   Queens: Flushing, Bisky.
  P. Pennsylvanicum L.   f. 1325.
   Frequent throughout the island.
  P. Persicaria L.   f. 1327.
   Frequent throughout the island.
  P. Careyi Olney.   f. 1329.
   Long Island, Hulst.   *Suffolk:* Miller & Young.
  P. hydropiperoides Mx.   f. 1332.
   Frequent throughout the island.
  P. Hydropiper L.   f. 1333.
   Frequent throughout the island.
  P. punctatum Ell.   f. 1334.
   Kings: Forbell's Landing, Hulst, Brooklyn, Brainerd.   *Suffolk:* Miller & Young.
  P. orientale L.   f. 1335.
   Frequent.   *Suffolk:* Miller & Young.
  P. Virginianum L.   f. 1336.
   Frequent thoughout the island.
  P. aviculare L.   f. 1337.
   Queens: Throughout, Bisky.   *Suffolk:* Greenport, Tillinghast; Miller & Young.

P. maritimum L. f. 1340.
  Queens: Rockaway Beach, Hulst ; N. Hempstead, Bisky. *Suffolk:* Miller & Young.
P. erectum L. f. 1342.
  Kings: New Lots, Brainerd ; Brooklyn, Hulst. *Queens:* Flushing, N. Hempstead ; Oyster Bay, Bisky. *Suffolk:* Miller & Young.
P. ramosissimum Mx. f. 1344.
  Queens: Flushing, Newtown, Bisky. *Suffolk:* Greenport, Tillinghast ; Miller & Young.
P. tenue Mx. f. 1346.
  Queens: Jamaica, Newtown, Oyster Bay, Bisky.
P. Convolvulus L. f. 1348.
  Frequent throughout the island.
P. cilinode Mx. f. 1349.
  Kings: Forbell's Landing, Hulst. *Queens:* Flushing, Newtown, Bisky.
P. dumetorum scandens (L.) Gray=*P. scandens* L. f. 1351.
  Frequent throughout the island.
P. Zuccarinii Small.
  Frequently cultivated.
P. sagittatum L. f 1354.
  Frequent throughout the island.
P. arifolium L. f. 1355.
  Queens: Maspeth, Hulst ; Flushing, Hempstead, N. Hempstead, Oyster Bay, Bisky. *Suffolk:* Greenport, Tillinghast; Miller & Young.

POLYGONELLA Mx.
P. articulatum (L.) Meissn. f. 1356.
  Frequent throughout the island.

CHENOPODIACEAE.
CHENOPODIUM L.
C. album L. f. 1359.
  Frequent throughout the island.
C. glaucum L. f. 1360.
  Kings: Brooklyn. *Queens:* Hicksville, Hulst, N. Hempstead, Bisky.
C. urbicum L. f. 1366.
  Kings: Brooklyn, Hulst. *Queens:* Flushing, Hempstead, Jamaica, Oyster Bay, Bisky. *Suffolk:* Miller & Young.
C. urbicum rhombifolium (Muhl.) Mog.
  Queens: Flushing, Oyster Bay, Bisky.

C. murale L. f. 1367.
: Kings:* Calverley, notes. *Queens:* Flushing, Oyster Bay, Bisky.

C. hybridum L. f. 1368.
: Long Island, Hulst. *Queens:* Flushing, Bisky.

C. rubrum L. f. 1369.
: *Kings:* Brooklyn, Hulst; Fort Hamilton, Brainerd. *Queens:* Glen Cove.

C. Bonus-Henricus L. f. 1370.
: *Queens:* Glen Cove, Bisky. *Suffolk:* Miller & Young.

C. Botrys L. f. 1371.
: *Kings:* Brooklyn, Hulst. *Queens:* Calverley. *Suffolk:* Miller & Young.

C. ambrosioides L. f. 1372.
: Frequent throughout the island.

C. anthelminticum L. f. 1373.
: Frequent throughout the island.

ATRIPLEX L.

A. hastata L. f. 1379.
: *Kings:* Calverley, notes; Fort Hamilton, Brainerd. *Queens:* Maspeth, Hulst; Flushing, Hempstead, N. Hempstead, Oyster Bay, Bisky. *Suffolk:* Miller & Young.

A. hastata littoralis (L.) Gray.
: *Queens:* Flushing, Bisky.

A. arenaria Nutt. f. 1383.
: *Queens:* Rockaway Beach, Hulst, Hempstead, N. Hempstead, Oyster Bay, Bisky. *Suffolk:* Sag Harbor, Mrs. L. D. Pychowska.

SALICORNIA L.

S. herbacea L. f. 1389.
: Frequent throughout the island.

S. Bigelovii Torr. f. 1390.
: Frequent throughout the island.

S. ambigua Michx. f. 1391.
: Frequent throughout the island.

DONDIA Adans.

D. Americana (Pers.) Britton. f. 1393.
: Frequent throughout the island.

D. maritima (L.) Druce. f. 1394.
: Frequent throughout.

SALSOLA L.

S. Kali L. f. 1396.
: Common throughout the island.

AMARANTHACEAE.
AMARANTHUS L.
- **A. retroflexus** L. f. 1398.
  Long Island, Hulst. *Suffolk:* Miller & Young.
- **A. hybridus** L. f. 1399.
  Kings: Gowanus, New Lots, Brainerd; Calverly, notes. *Queens:* Throughout, Bisky.
- **A. hybridus paniculatus** (L.) Uline and Bray.
  Suffolk: Miller & Young.
- **A. spinosus** L. f. 1400.
  Kings: Calverley, notes.
- **A. graecizans** L. f. 1402.
  Frequent throughout the island.
- **A. pumilus** Raf. f. 1408.
  Long Island, Hulst. *Suffolk:* Miller & Young.

ACNIDA L.
- **A. Cannabina** L. f. 1409.
  Kings: Calverley, notes. Forbell's Landing, Hulst. *Queens:* Flushing, Bisky. *Suffolk:* Miller & Young.
- **A. tamariscina** (Nutt.) Wood. f. 1410.
  Kings: Calverley, notes.

PHYTOLACCACEAE.
PHYTOLACCA L.
- **P. decandra** L. f. 1415.
  Common throughout the island.

AIZOACEAE.
SESUVIUM L.
- **S. maritimum** (Walt.) B.S.P. f. 1424.
  Suffolk: East Hampton, Brainerd, Miller & Young.

MOLLUGA L.
- **M. verticillata** L. f. 1425.
  Common throughout the island.

PORTULACACEAE.
CLAYTONIA L.
- **C. Virginica** L. f. 1429.
  Common throughout the island.

PORTULACA L.
- **P. oleracea** L. f. 1434.
  Common throughout the island.
- **P. grandiflora** Hook. f. 1437.
  Common throughout the island.

CARYOPHYLLACEAE.

AGROSTEMMA L.
- **A. Githago** (L.) Lam. f. 1438.
  Kings: New Lots. *Queens:* Cold Spring, Hulst; Flushing, N. Hempstead, Bisky. *Suffolk:* Greenport, Tillinghast; Miller & Young.

SILENE L.
- **S. stellata** (L.) Ait. f. 1441.
  Common throughout the island.
- **S. vulgaris** (Moench.) Garcke. f. 1443.
  Queens: Aqueduct, Hulst; Lawrence Station, Bisky. *Suffolk:* Sag Harbor, Mrs. L. D. Pychowska.
- **S. Caroliniana** Walt. f. 1448.
  Kings: Cypress Hill, Hulst. *Queens:* Flushing, Hempstead, N. Hempstead, Oyster Bay, Bisky. *Suffolk:* Shelter Island, Tillinghast; Miller & Young.
- **S. antirrhina** L. f. 1449.
  Frequent throughout the island.
- **S. Armeria** L. f. 1450.
  Kings: Cypress Hills, Hulst; Brooklyn, Brainerd.
- **S. noctiflora** L. f. 1451.
  Queens: Cold Spring, Maspeth, Hulst; Flushing, Bisky. *Suffolk:* Miller & Young.

LYCHNIS L.
- **L. alba** Mill. f. 1455.
  Queens: Cold Spring, Hulst; Flushing, N. Hempstead, Bisky.

TUNICA Adans.
- **T. Saxifraga** (L.) Scop. f. 1465.
  Queens: Flushing, J. Schrenk.

SAPONARIA L.
- **S. officinalis** L. f. 1466.
  Frequent throughout the island.

VACCARIA L.
- **V. Vaccaria** (L.) Britton. f. 1467.
  Queens: Far Rockaway, Hulst; Newtown, Ruger. *Suffolk:* Miller & Young.

DIANTHUS L.
- **D. prolifer** L. f. 1468.
  Suffolk: Sag Harbor, Mrs. L. D. Pychowska.
- **D. armeria** L. f. 1469.
  Common throughout the island.

ALSINE L.
  A. media L. f. 1475.
    Kings: Calverley; Forbell's, Torrey Club. *Queens:* Throughout, Bisky. *Suffolk:* Greenport, Tillinghast; Miller & Young.
  A. Holostea (L.) Britton. f. 1477.
    Queens: Newtown, Bisky; Ruger.
  A. longifolia (Muhl.) Britton. f. 1478.
    Queens: Aqueduct, Hulst; Flushing, Hempstead, N. Hempstead, Bisky. *Suffolk:* Miller & Young.
  A. graminea L. f. 1479.
    Kings: Fresh Pond, Hulst; Flushing, Bisky.

CERASTIUM L.
  C. viscosum L. f. 1484.
    Common throughout the island.
  C. vulgatum L. f. 1486.
    Kings: Hulst. *Queens:* Plandome, Eddy.
  C. arvense L. f. 1489.
    Common throughout the island.
  C. arvense oblongifolium (Torr.) Britt. & Hollick.
    Kings: Prospect Park.

SAGINA L.
  S. procumbens L. f. 1494.
    Kings: Cypress Hill, Hulst. *Queens:* Cold Spring, Hulst, Hempstead, Leggett, Bisky. *Suffolk:* Miller & Young.
  S. apetala Ard. f. 1495.
    Queens: Rockaway, Schrenk, Hempstead, Bisky.

ARENARIA L.
  A. serpyllifolia L. f. 1499.
    Kings: Forbell's Landing, New Lots, Hulst. *Queens:* Flushing, Hempstead, N. Hempstead, Bisky. *Suffolk:* Greenport, Tillinghast; Miller & Young.
  A. Caroliniana Walt. f. 1505.
    Kings: New Lots, Brainerd. *Suffolk:* East Hampton, Mrs. L. D. Pychowska; Miller & Young.

MOEHRINGIA L.
  M. lateriflora (L.) Fenzl. f. 1510.
    Common throughout the island.

AMMODENIA J. G. Gmel.
  A. peploides (L.) Rupr. f. 1512.
    Common on the sea beaches throughout the south coast of the island. *Queens:* Hempstead, Bisky. *Suffolk:* E. Hampton, Mrs. L. D. Pychowska; Miller & Young.

SPERGULA L.
  S. arvensis L. f. 1513.
    Kings: New Lots, Hulst. *Queens:* Cold Spring, Hulst. Centreville, Rudkin, Jamaica, Bisky. *Suffolk:* Miller & Young.

TISSA Adans.
  T. marina (L.) Britt. f. 1514.
    Kings: Forbell's Landing, Hulst. *Queens:* Aqueduct, Hulst; Rockaway Beach, Brainerd; Flushing, N. Hempstead, Newtown, Oyster Bay, Bisky. *Suffolk:* Wading River, Brainerd; Miller & Young.
  T. rubra (L.) Britt. f. 1516.
    Kings: New Lots, Newtown, Hulst. *Suffolk:* East Hampton, Mrs. L. D. Pychowska.

ANYCHIA Mx.
  A. dichotoma Mx. f. 1522.
    Queens: Richmond Hill, Cold Spring, Hulst; Flushing, N. Hempstead, Oyster Bay, Bisky. *Suffolk:* Wading River. Brainerd.

SCLERANTHUS L.
  S. annuus L. f. 1524.
    Frequent throughout the island.

NYMPHACEAE.

BRASENIA Schreb.
  B. purpurea (Michx.) Casp. f. 1526.
    Queens: Flushing, Hulst. *Suffolk:* Miller & Young.

NYMPHAEA L.
  N. advena Soland. f. 1527.
    Common throughout the island.

CASTALIA Salisb.
  C. odorata (Dryand) Woodr. & Wood. f. 1531.
    Common throughout the island.

NELUMBO Adans.
  N. lutea (Willd.) Pers. f. 1534.
    Frequent throughout the island.

CERATOPHYLLACEAE.

CERATOPHYLLUM L.
  C. demersum L. f. 1536.
    Queens: Cold Spring, Hulst.

MAGNOLIACEAE.
MAGNOLIA L.
  M. tripetala L.   f. 1539.
  Kings: Prospect Park.
  M Virginiana L.   f. 1540.
  Suffolk: M. H. Delafield; Speonk, Rusby.

LIRIODENDRON L.
  L. Tulipifera L.   f. 1542.
  Common throughout the island.

RANUNCULACEAE.
CALTHA L.
  C. palustus L.   f. 1545.
  Common throughout the island.

TROLLIUS L.
  T. laxus Salisb.   f. 1548.
  Queens: Woodhaven, Miss M. O. Steel.

HELLEBORUS L.
  H. viridis L.   f. 1549.
  Has been reported wild on Long Island, Gray's Manual, 1890, p. 45. *Queens:* Glen Cove, I. Coles, Jamaica, Oyster Bay, Bisky.

XANTHORRHIZA L'Her.
  X. apiifolia L'Her.   f. 1553.
  Kings: Prospect Park.   Probably an introduced species.

ACTAEA L.
  A. rubra (Ait.) Willd.   f. 1554.
  Queens: Newtown, Oyster Bay, Bisky.
  A. alba (L.) Mill.   f. 1555.
  Queens: Hulst, Richmond Hill; Flushing, Newtown, Oyster Bay, Bisky.

CIMICIFUGA L.
  C racemosa (L.) Nutt.   f. 1556.
  Queens: Woodhaven, Hulst; Flushing, Jamaica, N. Hempstead, Oyster Bay, Bisky.

AQUILEGIA L.
  A. Canadensis L.   f. 1559.
  Kings: Prospect Park.   *Queens:* Hulst, Glen Cove, Coles; Oyster Bay, Bisky.   *Suffolk:* Miller & Young.
  A. vulgaris L.   f. 1561.
  Kings: Hulst.   *Suffolk:* Miller & Young.

DELPHINIUM L.
  D. Consolida L.   f. 1562.
  Kings: Hulst.

ANEMONE L.
  A. Virginiana L. f. 1573.
    Queens: Richmond Hill, Calverley; Flushing, Newtown, N. Hempstead, Oyster Bay, Bisky. *Suffolk:* Miller & Young, East Hampton, Mrs. L. D. Pychowska.
  A. quinquefolia L. f. 1576.
    Common throughout the island.

HEPATICA Scop.
  H. Hepatica (L.) Karst. f. 1578.
    Common throughout the island.

SYNDESMON Hoffmg.
  S. thalictroides (L.) Hoffmg. f. 1580.
    Common throughout the island.

CLEMATIS L.
  C. Virginiana L. f. 1582.
    Common throughout the island.
  C. ochroleuca Ait. f. 1588.
    Kings: Brooklyn, Britton & Brown. *Queens:* Plandome, Eddy.

RANUNCULUS L.
  R. delphinifolius Torr. f. 1595.
    Kings: Hulst, Brainerd. *Queens:* Ridgewood, M. Ruger. *Suffolk:* Greenport, Tillinghast.
  R. reptans L. f. 1602.
    Queens: Calverley, Hulst.
  R. abortivus L. f. 1609.
    Common throughout the island.
  R. sceleratus L. f. 1612.
    Queens: Calverley, Hulst; Flushing, Newtown, N. Hempstead, Oyster Bay, Bisky. *Suffolk:* Greenport, Tillinghast, Miller & Young.
  R. recurvatus Poir. f. 1613.
    Common throughout the island.
  R. acris L. f. 1614.
    Common throughout the island.
  R. bulbosus L. f. 1615.
    Common throughout the island.
  R. Pennsylvanicus L. f. 1616.
    Kings: Hulst. *Queens:* Newtown, M. Ruger. *Suffolk:* Miller & Young.
  R. repens L. f. 1618.
    Kings: Hulst. *Queens:* Richmond Hill, Hulst; Flushing, N. Hempstead, Bisky.

R. fascicularis Muhl. f. 1621.
   Common throughout the island.
BATRACHIUM S. F. Gray.
  B. trichophyllum (Chaix) Bossch. f. 1626.
    Kings: New Lots, Brainerd. *Queens:* Jamaica, F. T. Allen. *Suffolk:* Miller & Young.
FICARIA Huds.
  F. Ficaria (L.) Karst. f. 1629.
    Queens: Flushing, Bisky.
OXYGRAPHIS Burge.
  O. Cymbalaria (Pursh) Prantl. f. 1631.
    Queens: Hempstead, W. H. Leggett. *Suffolk:* East Hampton, Mrs. L. D. Pychowska; Miller & Young.
THALICTRUM L.
  T. dioicum L. f. 1634.
    Kings: Hulst, Flatbush. *Queens:* Calverley; Newtown, Bisky. *Suffolk:* Greenport, Tillinghast; Miller & Young.
  T. purpurascens L. f. 1637.
    Queens: Newtown, Bisky. *Suffolk:* Greenport, Tillinghast; Miller & Young.
  T. polygamum Muhl. f. 1638.
    Common throughout the island.

BERBERIDACEAE.
BERBERIS L.
  B. vulgaris L. f. 1640.
    Common throughout the island.

MENISPERMACEAE.
MENISPERMUM L.
  M. Canadense L. f. 1649.
    Queens: Aqueduct, Hulst; Flushing, Bisky.

CALYCANTHACEAE.
BUTNERA Duham.
  B. florida (L.) Kearney. f. 1650.
    In cultivation, Prospect Park.
  B. fertilis (Walt.) Kearney. f. 1651.
    In cultivation, Prospect Park.

LAURACEAE.
SASSAFRAS Nees & Eberm.
  S. Sassafras (L.) Karst. f. 1654.
    Frequent throughout the island.

BENZOIN Fabric.
  B. Benzoin (L.) Coulter. f. 1656.
    Frequent throughout the island.

PAPAVERACEAE.

ARGEMONE L.
  A. Mexicana L. f. 1663.
    Kings: Brainerd. *Queens:* Flushing, Bisky.

SANGUINARIA L.
  S. Canadensis L. f. 1665.
    Common throughout the island.

GLAUCIUM Juss.
  G. Glaucium (L.) Karst. f. 1667.
    Queens: Newtown, Bisky; Long Island City, Addison Brown.

CHELIDONIUM L.
  C. majus L. f. 1668.
    Common throughout the island.

CRUCIFERAE.

LEPIDIUM L.
  L. campestre (L.) R.Br. f. 1684.
    Common throughout the island.
  L. Draba L. f. 1685.
    Queens: Astoria, D. C. Eaton (station destroyed).
  L. ruderale L. f. 1686.
    Kings: Hulst.
  L. Virginicum L. f. 1687.
    Common throughout the island.

CORONOPUS Gaertn.
  C. didymus (L.) J. E. Smith. f. 1690.
    Kings: Brooklyn, Hulst.

THLASPI L.
  T. arvense L. f. 1692.
    Queens: Cold Spring Harbor, Hulst.

SISYMBRIUM L.
  S. officinale (L.) Scop. f. 1696.
    Common throughout the island.

CAKILE Gaertn.
  C. edentula (Bigel.) Hook. f. 1699.
    Common throughout.

SINAPIS L.
  S. alba L. f. 1700.
    Escaped.

BRASSICA L.
  B. nigra (L.) Koch. f. 1701.
    Frequent throughout.
  B. arvensis (L.) B.S.P. f. 1703.
    Frequent.
  B. campestris L. f. 1704.
    Common throughout the island.

DIPLOTAXIS DC.
  D. tenuifolia (L.) DC. f. 1705.
    Kings: Brooklyn, Hulst.

RAPHANUS L.
  R. Raphanistrum L. f. 1707.
    Common throughout the island.
  R. sativus L. f. 1708.
    Common throughout the island.

BARBAREA R.Br.
  B. Barbarea (L.) MacM. f. 1709.
    Kings: New Lots, Hulst. *Queens:* Flushing, Newtown, N. Hempstead, Bisky. *Suffolk:* Greenport, Tillinghast; Miller & Young.
  B. praecox (Smith) R.Br. f. 1711.
    Queens: Cold Spring, Hulst; Parson's nursery, Leggett.

RORIPA Scop.
  R. sylvestris (L.) Bess. f. 1713.
    Kings: Brooklyn, Hulst. *Queens:* Flushing Bay, Bisky.
  R. palustris (L.) Bess. f. 1717.
    Kings: Cypress Hills, Hulst. *Queens:* Newtown, Bisky. *Suffolk:* Miller & Young.
  R. hispida (Desv.) Britt. f. 1718.
    Kings: Cypress Hills, Hulst. *Queens. Suffolk:* Miller & Young.
  R. Nasturtium (L.) Rusby. f. 1721.
    Kings: Hulst. *Queens:* Flushing, Newtown, N. Hempstead, Oyster Bay, Bisky. *Suffolk:* Wading River, Brainerd, Miller & Young; E. Hampton, Mrs. L. D. Pychowska.
  R. Armoracia (L.) A. S. Hitchcock. f. 1722.
    Frequent throughout the island.

CARDAMINE L.
  C. hirsuta L. f. 1725.
    Kings: Brainerd; Forbell's Landing, Cypress Hills, Hulst. *Queens:* Hulst; Flushing, N. Hempstead, Oyster Bay, Bisky.
  C. rotundifolia Mx. f. 1734.
    Queens: Pardegot, Brainerd.

DENTARIA L.
- **D. laciniata** Muhl. f. 1735.
  Long Island, Chas. H. Hall. *Queens:* Flushing, Bisky.
- **D. diphylla** Mx. f. 1736.
  Long Island, Chas. H. Hall. *Queens*: Flushing, Bisky, R. Lawrence.
- **D. maxima** Nutt. f. 1737.
  Queens: Newtown, Bisky; Bowery Bay, M. Rugers.

BURSA Weber.
- **B. Bursa-pastoris** (L.) Britton. f. 1752.
  Common throughout the island.

CAMELINA Crantz.
- **C. sativa** (L.) Crantz. f. 1753.
  Suffolk: Miller & Young.

DRABA L.
- **D. verna** L. f. 1755.
  Common throughout the island.
- **D. Caroliniana** Walt. f. 1756.
  Queens: Ridgewood Aqueduct, W. H. Rudkin.

ARABIS L.
- **A. lyrata** L. f. 1772.
  Kings: Calverley; Forbell's Landing. *Queens:* Richmond Hill, Woodhaven, Hulst; Flushing, Oyster Bay, Bisky. *Suffolk:* Southhold, Tillinghast; Miller & Young.
- **A. laevigata** (Muhl.) Poir. f. 1778.
  Queens: Richmond Hill, in Herb. Poggenberg.
- **A. Canadensis** L. f. 1779.
  Kings: Cypress Hills, Hulst. *Queens:* Jamaica, Brainerd; Richmond Hill, Hulst; Newtown, Bisky.
- **A. glabra** (L.) Bernh. f. 1781.
  Suffolk: Miller & Young.

KONIGA Adans.
- **K. maritima** (L.) R.Br. f. 1788.
  Suffolk: Miller & Young.

HESPERIS L.
- **H. matronalis** L. f. 1790.
  Queens: Hulst; Jamaica, W. H. Rudkin.

CAPPARIDACEAE.

CLEOME L.
- **C. spinosa** L. f. 1792.
  Queens: Hulst; Torrey Club.

POLANISIA Raf.
  P. graveolens Raf.   f. 1796.
    Queens: Centreville, W. H. Rudkin.

RESEDACEAE.
RESEDA L.
  R. Luteola L.   f. 1798.
    Suffolk: Greenport, Tillinghast; Miller & Young.

SARRACENIACEAE.
SARRACENIA L.
  S. purpurea L.,   f. 1801.
    Queens: Hulst, Richmond Hill, south of Jerusalem, Coles; Oyster Bay, Bisky. *Suffolk:* Miller & Young.

DROSERACEAE.
DROSERA L.
  D. rotundifolia L.   f. 1803.
    Kings: Forbell's Landing, Hulst; Greenwood, Brainerd. *Suffolk:* Sag Harbor, Mrs. S. L. Zabriskie; Napeague Beach, Mrs. L. D. Pychowska; Miller & Young.
  D. intermedia Hayne.   f. 1804.
    Kings: Cedarhurst, Hulst. *Suffolk:* Miller & Young.
  D. filiformis Raf.   f. 1807.
    Kings: Greenwood, Brainerd. *Suffolk:* Sag Harbor, Mrs. S. L. Zabriskie; Napeague Beach, Mrs. L. D. Pychowska; Miller & Young.

CRASSULACEAE.
SEDUM L.
  S. Telephium L.   f. 1811.
    Common throughout the island.
  S. acre L.   f. 1813.
    Common throughout the island. Escaped from gardens.
  S. ternatum Mx.   f. 1818.
    Queens: West Flushing, M. Rugers.
PENTHORUM L.
  P. sedoides L.   f. 1821.
    Long Island, Hulst. *Queens:* Flushing, Hempstead; Jamaica, Oyster Bay, Bisky. *Suffolk:* Miller and Young.

SAXIFRAGACEAE.
SAXIFRAGA L.
  S. Pennsylvanica L.   f. 1831.
    Kings: Cypress Hills, Hulst. *Queens:* Richmond Hill, Hulst; Glen Cove, Coles.

S. Virginiensis Mx. f. 1833.
Common throughout the island.

HEUCHERA L.
H. Americana L. f. 1845.
Queens: Richmond Hill, Hulst ; Flushing, Oyster Bay, Bisky.

MITELLA L.
M. diphylla L. f. 1848.
Queens: Richmond Hill, Hulst.

CHRYSOSPLENIUM L.
C. Americanum L. f. 1850.
Queens: Cold Spring, Hulst ; Flushing, Bisky. *Suffolk:* Miller & Young.

PARNASSIA L.
P. Caroliniana Mx. f. 1852.
Queens: Maspeth, Hulst; Flushing, Hempstead, Newtown, Bisky; Woodside, A. Brown. *Suffolk:* Good ground.

GROSSULARIACEAE.

RIBES L.
R. Cynosbati L. f. 1865.
Queens: Woodside, Hulst.
R. oxyacanthoides L. f. 1868.
Queens: Flushing, Bisky. *Suffolk:* Greenport, Tillinghast ; Miller & Young.
R. rotundifolium Mx. f. 1869.
Long Island, Hulst.
R. floridum L'Her. f. 1874.
Queens: Rugers. ? *Suffolk:* Miller & Young.
R. rubrum L. f. 1875.
Common throughout the island.
R. aureum Pursh. f. 1877.
Kings: Prospect Park.

HAMAMELIDACEAE.

HAMAMELIS L.
H. Virginiana L f. 1879.
Common throughout the island.

LIQUIDAMBAR L.
L. Styraciflua L. f. 1880.
Common throughout the island.

PLATANACEAE.

PLATANUS L.
P. occidentalis L. f. 1881.
Frequent throughout the island.

ROSACEAE.

OPULASTER Medic.
  O. opulifolius (L.) Kuntze. f. 1882.
    Kings: Forbell's Landing, Hulst.

SPIRAEA L.
  S. salicifolia L. f. 1883.
    Kings: Cypress Hills, Hulst ; Prospect Park ; Forbell's Landing, Hulst. *Queens:* Richmond Hill ; Flushing, Newtown, Bisky. *Suffolk:* Miller & Young.
  S. tomentosa L. f. 1884.
    Kings: Cypress Hills, Forbell's Landing, Hulst ; New Lots. *Queens:* Richmond Hill ; Flushing, Hempstead, Jamaica, Oyster Bay, Bisky. *Suffolk:* Fisher's Island, Tillinghast ; Miller & Young.
  S. corymbosa Raf. f. 1885.
    Long Island, Hulst ?

RUBUS L.
  R. odoratus L. f. 1890.
    Common throughout the eastern end of the island.
  R. strigosus Mx. f. 1894.
    Common throughout the island.
  R. occidentalis L. f. 1896.
    Queens: Flushing, N. Hempstead, Oyster Bay, Bisky ; Glen Cove. *Suffolk:* Miller & Young.
  R. villosus Ait. f. 1898.
    Common throughout the island.
  R. cuneifolius Pursh. f. 1901.
    Kings: Forbell's Landing, Hulst.
  R. hispidus L. f. 1902.
    Queens: Valley Stream ; Throughout, Bisky. *Suffolk:* Miller & Young.
  R Canadensis L. f. 1906.
    Common throughout the island.

FRAGARIA L.
  F. Virginiana Duchesne. f. 1908.
    Common throughout the island.
  F. vesca L. f. 1910.
    Common throughout the island.

DUCHESNEA J. E. Smith.
  D. Indica (Andr.) Focke. f. 1912.
    Queens: Glen Cove.

POTENTILLLA L.
- **P. argentea** L. f. 1914.
  Kings: Brooklyn, Hulst. *Queens:* Flushing, Hempstead, Newtown, Bisky. *Suffolk:* Greenport, Tillinghast; Miller & Young.
- **P. recta** L. f. 1917.
  Queens: Flushing, Oyster Bay, Bisky. *Suffolk:* Miller & Young.
- **P. Monspeliensis** L. f. 1922.
  Common throughout the island.
- **P. Anserina** L. f. 1934.
  Queens: Glen Cove, Coles. *Suffolk:* Greenport, Tillinghast; Miller & Young.
- **P. Canadensis** L. f. 1935.
  Common throughout the island.

GEUM L.
- **G. vernum** (Raf.) T. & G. f. 1943.
  Kings: Prospect Park, Mrs. L. LeBrun; Jelliffe; Long Island, Brainerd, 1866.
- **G. Canadense** Jacq. f. 1944.
  Queens: Sand's Point; Brainerd; Maspeth, Aqueduct, Hulst; Flushing, N. Hempstead, Oyster Bay, Bisky. *Suffolk:* Greenport, Tillinghast; East Hampton, Mrs. L. D. Pychowska; Miller & Young.
- **G. Virginianum** L. f. 1945.
  Kings: New Lots, Brainerd. *Queens:* Flushing, Bisky. *Suffolk:* Miller & Young.
- **G. strictum** Ait. f. 1948.
  Queens: Newtown, Bisky.

AGRIMONIA L.
- **A. hirsuta** (Muhl.) Bicknell. f. 1957.
  Frequent throughout the island.
- **A. mollis** (T. & G.) Britton. f. 1960.
  Frequent.
- **A. parviflora** Soland. f. 1962.
  Queens: Cold Spring, Hulst. *Suffolk:* Pardegot, Brainerd.

SANGUISORBA L.
- **S. Canadensis** L. f. 1964.
  Common throughout the island.

ROSA L.
- **R. blanda** Ait. f. 1966.
  Queens: Jamaica, Bisky. *Suffolk:* Miller & Young.
- **R. Carolina** L. f. 1970.
  Common throughout the island.

R. humilis Marsh. f. 1971.
　　Kings: Cypress Hills, Hulst. *Queens:* Flushing, Hempstead, Jamaica, Oyster Bay, Bisky; Flushing, Schenk. *Suffolk:* Miller & Young.

R. canina L. f. 1973.
　　Queens: Cold Spring Harbor, Hulst.

R. rubiginosa L. f. 1974.
　　Kings: Forbell's Landing, Hulst. *Queens:* Jamaica, Hulst; Flushing, Hempstead, N. Hempstead, Oyster Bay, Bisky; Sand's Point, W. H. Leggett. *Suffolk:* Greenport, Tillinghast; Miller & Young.

POMACEAE.

SORBUS L.
　S. Americana Marsh. f. 1975.
　　Prospect Park.

MALUS Juss.
　M. Malus L. f. 1982.
　　Common throughout the island.

ARONIA Pers.
　A. arbutifolia (L.) Ell. f. 1983.
　　Common throughout the island.
　A. nigra (Willd.) Britton. f. 1984.
　　Queens: Flushing, Hempstead; Jamaica, Bisky. *Suffolk:* Miller & Young.

AMELANCHIER Medic.
　A. Canadense (L.) Medic. f. 1985.
　　Long Island, Hulst. *Queens:* Flushing, Oyster Bay, Bisky, *Suffolk:* Miller & Young; E. Hampton, Mrs. L. D. Pychowska.
　A. Canadense obovalis (Mx.) Torr.
　　Suffolk: Greenport, Tillinghast; Miller & Young.
　A. rotundifolia (Mx.) Roem. f. 1988.
　　Queens: Hicksville, Hulst.

CRATAEGUS L.
　C. Crus-Galli L. f. 1991.
　　Queens: Flushing, Bisky. *Suffolk:* Miller & Young.
　C. Oxyacantha L. f. 1995.
　　Kings: Prospect Park. *Suffolk:* Miller & Young.
　C. coccinea'L. f. 1998.
　　Kings: Prospect Park. *Queens:* Woodhaven, Torrey Club, 1891; Flushing, Newtown, Bisky. *Suffolk:* Miller & Young.
　C. uniflora Muench. f. 2004.
　　Suffolk: Miller & Young.

DRUPACEAE.

PRUNUS L.
- **P. Americana** Marsh.   f. 2007.
  Suffolk: Miller & Young.
- **P. maritima** Wang.   f. 2013.
  Kings: Coney Island, Brainerd.   *Queens:* Rockaway Beach; Cold Spring Harbor, Flushing, Hempstead, Newtown, Oyster Bay, Bisky.   *Suffolk:* East Hampton, Mrs. L. D. Pychowska; Miller & Young.
- **P. spinosa** L.   f. 2016.
  Suffolk: Miller & Young.
- **P. Cerasus** L.   f. 2020.
  Common throughout the island.
- **P. Pennsylvanica** L. f.   f. 2022.
  Queens: Oyster Bay, Bisky; Glen Cove, Coles.
- **P. Mahaleb** L.   f. 2023.
  Suffolk: East Hampton, M. L. Delafeld, Jr.
- **P. Virginiana** L.   f. 2024.
  Common throughout the island.
- **P. serotina** L.   f. 2026.
  Common throughout the island.

AMYGDALUS L.
- **A. Persica** L.   f. 2027.
  Escaped from cultivation.

CAESALPINACEAE.

CERCIS L.
- **C. Canadensis** L.   f. 2033.
  Kings: Prospect Park.

CASSIA L.
- **C. nictitans** L.   f. 2034.
  Kings: Gowanus, Brainerd; Forbell's Landing, Hulst; Cedarhurst, Torrey Club.   *Queens:* Flushing, Hempstead, Oyster Bay, Bisky.   *Suffolk:* Miller & Young.
- **C. Chamaecrista** L.   f. 2035.
  Common throughout the island.
- **C. Marylandica** L.   f. 2037.
  Kings: Gowanus, Brainerd; Cedarhurst, Torrey Club; Forbell's Landing, Hulst.   *Queens:* Flushing, Hempstead, Oyster Bay, Bisky.   *Suffolk:* Sag Harbor, Mrs. L. D. Pychowska.

GLEDITSIA L.
- **G. triacanthos** L.   f. 2041.
  Kings: Cypress Hills, Hulst; Prospect Park.   *Queens:* Hicksville, Hulst; Flushing, Oyster Bay, Bisky.

GYMNOCLADUS Lam.
   G. dioica (L.) Koch. f. 2043.
     Kings: Prospect Park.

PAPILIONACEAE.

CLADRASTIS Raf.
   C. lutea (Mx.) Koch. f. 2046.
     Kings: Prospect Park.

BAPTISIA Vent.
   B. tinctoria (L.) R. Br. f. 2050.
     Common throughout the island.

CROTALARIA L.
   C. sagittalis L. f. 2055.
     Kings: Cypress Hills, Hulst; Gowanus, Brainerd. *Queens:* Maspeth, Hulst; Flushing, Bisky. *Suffolk:* Greenport, Tillinghast; Miller & Young; East Hampton, Mrs. L. D. Pychowska.

LUPINUS L.
   L. perennis L. f. 2057.
     Queens: Woodhaven, Hicksville, Hulst; Jamaica, Oyster Bay. Bisky; Plandome, Eddy. *Suffolk:* East Hampton, Mrs. L. D. Pychowska; Miller & Young; Yaphank, Brainerd.

GENISTA L.
   G. tinctoria L. f. 2062.
     Queens: Richmond Hill, Hulst.

CYTISUS L.
   C. scoparius (L.) Link. f. 2063.
     Queens: Richmond Hill.

MEDICAGO L.
   M. sativa L. f. 2064.
     Kings: New Lots, Hulst. *Queens:* Flushing, Bisky. *Suffolk:* Miller & Young; Sag Harbor, Mrs. L. D. Pychowska.
   M. lupulina L. f. 2065.
     Kings: Hulst. *Queens:* Northville, Brainerd; Plandome, Eddy. *Suffolk:* Greenport, Tillinghast; Miller & Young.

MELILOTUS Juss.
   M. alba Desv. f. 2068.
     Common throughout the island.
   M. officinalis (L.) Lam. f. 2069.
     Frequent throughout.

TRIFOLIUM L.
 T. agrarium L.. f. 2070.
  Kings: Forbell's Landing, Hulst; Brooklyn. *Queens:* Richmond Hill; throughout, Bisky. *Suffolk:* Greenport, Tillinghast; Miller & Young.
 T. procumbens L.. f. 2071.
  Common throughout the island.
 T. incarnatum L.. f. 2073.
  Suffolk: Shelter Island, Hulst.
 T. arvense L.. f. 2074.
  Kings: Forbell's, Hulst. *Queens:* Woodhaven, throughout, Bisky. *Suffolk:* Greenport, Tillinghast; Miller & Young; East Hampton, Mrs. L. D. Pychowska.
 T. pratense L.. f. 2075.
  Common throughout the island.
 T. hybridum L.. f. 2081.
  Kings: Brooklyn, New Lots, Hulst. *Queens:* Jamaica, Hulst; Hempstead, Bisky.
 T. repens L.. f. 2083.
  Common throughout the island.

AMORPHA L..
 A. fruticosa L.. f. 2101.
  Kings: Hulst, Prospect Park. *Queens:* Flushing, Brainerd. *Suffolk:* Miller & Young.

CRACCA L..
 C. Virginiana L.. f. 2117.
  Queens: Selden, Brainerd; Hicksville, Woodlawn, Hulst. *Suffolk:* Miller & Young; East Hampton, Mrs. L. D. Pychowska.

ROBINIA L..
 R. Pseudacacia L.. 2121.
  Common throughout the island.
 R. viscosa Vent. f. 2122.
  Queens: Flushing, Bisky. *Suffolk:* Miller & Young.
 R. hispida L. f. 2123.
  Long Island, Hulst, cultivated.

CORONILLA L..
 C. varia L.. f. 2164.
  Suffolk: Bellport, Charles H. Hall; Shelter Island, Hulst.

STYLOSANTHES Sw.
 S. biflora (L.) B.S P. f. 2168..
  Queens: Selden, Brainerd; Oyster Bay, Bisky. Glen Cove, Coles. *Suffolk:* Miller & Young.

MEIBOMIA Adans.
  M. nudiflora (L.) DC. f. 2170.
    Kings: Cypress Hills, Hulst ; Prospect Park. *Queens:* Hulst ; Flushing, Jamaica, Oyster Bay, Bisky. *Suffolk:* Greenport, Tillinghast ; Miller & Young ; Sag Harbor, Mrs. L. D. Pychowska.
  M. grandiflora (Walt.) DC. f. 2171.
    Kings: Prospect Park. *Queens:* Flushing, N. Hempstead, Oyster Bay, Bisky.
  M. Michauxii Vail. f. 2174.
    Kings: Cypress Hills, Hulst. *Queens:* Centreville, Rudkin. *Suffolk:* Miller & Young.
  M. canescens (L.) Kuntze. f. 2179.
    Queens: College Point, Rudkin.
  M. bracteosa (Mx.) Kuntze. f. 2181.
    Kings: Cypress Hills, Hulst. *Queens:* Centreville, Rudkin. *Suffolk:* Sag Harbor, Mrs. L. D. Pychowska.
  M. paniculata (L.) Kuntze. f. 2182.
    Kings: Cypress Hills, Hulst. *Queens:* Flushing, N. Hempstead, Oyster Bay, Bisky. *Suffolk:* Miller & Young.
  M. laevigata (Nutt.) Kuntze. f. 2183.
    Long Island. Hulst. *Suffolk:* Miller & Young.
  M. viridiflora Kuntze. f. 2185.
    Long Island, Hulst. *Queens:* Flushing, Newtown, Bisky.
  M. Dillenii (Darl.) Kuntze. f. 2186.
    Queens: Newtown, Bisky.
  M. Canadense (L.) Kuntze. f. 2188.
    Kings: Cypress Hills. Hulst.
  M. rigida (Ell.) Kuntze. f. 2189.
    Suffolk: Sag Harbor, Mrs. L. D. Pychowska ; Miller & Young.
  M. Marylandica (L.) Kuntze. f. 2190.
    Queens: Jamaica, Oyster Bay, Bisky. *Suffolk:* Sag Harbor, Mrs. L. D. Pychowska ; Miller & Young.
  M. obtusa (Muhl.) Vail. f. 2191.
    Kings: Cypress Hills, Hulst ; Cedarhurst, Torrey Club. *Queens:* Flushing, Hempstead, Jamaica, N. Hempstead, Bisky.

LESPEDEZA Michx.
  L. repens (L.) Bart. f. 2192.
    Queens: Woodhaven, Hulst ; Hempstead, Newtown, Oyster Bay, Bisky. *Suffolk:* East Hampton, Mrs. L. D. Pychowska ; W. E. Wheelock ; Miller & Young.
  L. procumbens Michx. f. 2193.
    Queens: Newtown, Bisky. *Suffolk:* Miller & Young.

L. violacea (L.) Pers. f. 2195.
   Queens: Richmond Hill, Hulst. *Suffolk:* East Hampton, Mrs. L. D. Pychowska; Miller & Young.

L. Stuvei Nutt. f. 2196.
   Suffolk: East Hampton, Mrs. L. D. Pychowska; W. E. Wheelock.

L. Virginica (L.) Britt. f. 2198.
   Queens: Throughout, Bisky. *Suffolk:* East Hampton, Mrs. L. D. Pychowska; Miller & Young.

L. hirta (L.) Ell. f. 2199.
   Common throughout the island.

L. capitata Mx. f. 2200.
   Common throughout the island.

L. angustifolia (Pursh) Ell. f. 2201.
   Kings: Forbell's Landing, Hulst.

VICIA L.
V. Cracca L. f. 2204.
   Queens: Maspeth, Hulst; Glen Cove, Rudkin.

V. tetrasperma (L.) Moench. f. 2210.
   Queens: Flushing, Bisky.

V. hirsuta (L.) Koch. f. 2211.
   Queens: Newtown, Bisky. *Suffolk:* Patchogue, Ruger.

V. sativa L. f. 2212.
   Kings: Prospect Park? *Queens:* Flushing, Newtown, Bisky. *Suffolk:* Greenport, Tillinghast; Miller & Young.

V. angustifolia Roth. f. 2213.
   Suffolk: Miller & Young.

LATHYRUS L.
L. maritimus (L.) Bigel. f. 2215.
   Common along the coast throughout the island.

L. palustris L. f. 2217.
   Queens: Flushing, Bisky. *Suffolk:* Miller & Young.

L. myrtifolius Muhl. f. 2218.
   Queens: St. Roman's Hill, Poggenburg. ?

CLITORIA L.
C. Mariana L. f. 2224.
   Kings: Brooklyn, Britton & Brown.

FALCATA Gmel.
F. comosa (L.) Kuntze. f. 2225.
   Common throughout the island.

APIOS Moench.
A. Apios (L.) MacM. f. 2227.
   Common throughout the island.

GALACTIA R. Br.
   G. volubilis (L.) Britton. f. 2229.
      Queens: Newtown, Bisky. *Suffolk:* Miller & Young.

PHASEOLUS L.
   P. polystachyus (L.) B.S.P. f. 2234.
      Suffolk: Shelter Island, Hulst.

STROPHOSTYLES Ell.
   S. helvola (L.) Britton. f. 2235.
      Kings: Bay Ridge. *Queens:* Rockaway Beach, Hulst; Throughout, Bisky. *Suffolk:* Greenport, Tillinghast; East Hampton, Mrs. L. D. Pychowska; Miller & Young.
   S. umbellata (Muhl.) Britton. f. 2236.
      Kings: Cedarhurst, Hulst; Brainerd; Bay Ridge. *Queens:* Hempstead, Bisky. *Suffolk:* East Hampton, Mrs. L. D. Pychowska; Miller & Young.

GERANIACEAE.

GERANIUM L.
   G. maculatum L. f. 2239.
      Common throughout the island.
   G. Robertianum L. f. 2240.
      Queens: Rockaway Beach, Hulst; Richmond Hill; Hempstead, Bisky.
   G. rotundifolium L. f. 2243.
      Queens: Long Island City, A. Brown; *fide* Bisky.
   G. Carolinianum L. f. 2244.
      Queens: Rockaway Beach, Hulst; Flushing, Bisky. *Suffolk:* Greenport, Tillinghast; Miller & Young.
   G. dissectum L. f. 2246.
      Kings: Brooklyn, Hulst.
   G. pusillum L. f. 2247.
      Queens: Hulst; Flushing, Newtown, Bisky.

ERODIUM L'Her.
   E. cicutarium (L.) L'Her. f. 2249.
      Queens: Sand's Point, Rev. Chas. H. Hall; Flushing, Oyster Bay, Bisky.

OXALIDACEAE.

OXALIS L.
   O. violacea L. f. 2251.
      Kings: Cypress Hills, Hulst; Prospect Park. *Queens:* Richmond Hill; Flushing, Oyster Bay, Bisky.

O. corniculata L. f. 2252.
: New Lots, Hulst ; Prospect Park, cultivated.
O. stricta L. f. 2254.
Kings: New Lots, Hulst ; Prospect Park. *Queens:* Richmond Hill ; throughout, Bisky. *Suffolk:* E. Hampton, Mrs. L. D. Pychowska ; Miller & Young.

LINACEAE.
LINUM L.
L. usitatissimum L. f. 2258.
Kings: Brooklyn. *Queens:* Flushing, Hempstead, Bisky. *Suffolk:* Miller & Young.
L. Virginianum L. f. 2260.
Kings: Cypress Hills, Hulst. *Queens:* Woodhaven, Hulst ; Richmond Hill ; Flushing, Hempstead, N. Hempstead, Bisky. *Suffolk:* E. Hampton, Mrs. L. D. Pychowska ; Miller & Young.
L. striatum Walt. f. 2263.
Queens: Centreville, M. Rugers. *Suffolk:* E. Hampton, Mrs. M. Pychowska ; Miller & Young.

RUTACEAE.
XANTHOXYLUM L.
X. Americanum Mill. f. 2269.
Queens: Oyster Bay, Bisky ; Glen Cove, Coles, Rudkin.
RUTA L.
R. graveolens L.
Queens: Oyster Bay, Bisky ; Glen Cove, Coles.
PTELEA L.
P. trifoliata L. f. 2271.
Suffolk: Miller & Young.

SIMARUBACEAE.
AILANTHUS Desf.
A. glandulosa Desf. f. 2272.
Common throughout the island.

POLYGALACEAE.
POLYGALA L.
P. lutea L. f. 2275.
Queens: Babylon ? Britton & Brown. *Suffolk:* Miller & Young.
P. cruciata L. 2276.
Kings: New Lots, Brainerd. *Queens:* Forbell's Landing, Hulst. *Suffolk:* East Hampton, Mrs. L. D. Pychowska ; Miller & Young.

P. verticellata L.. f. 2278.
Kings: Hulst. *Queens:* Hulst. *Suffolk:* Greenport. Tillinghast ; Miller & Young.

P. ambigua Nutt. f. 2279.
Kings: Cypress Hills, Hulst.

P. viridescens L.. f. 2281.
Kings: Forbell's Landing, Hulst. *Suffolk:* Greenport, Tillinghast ; East Hampton, Mrs. L. D. Pychowska.

P. Nuttallii T. & G. f. 2284.
Queens: Hicksville, Hulst. *Suffolk:* Greenport, Tillinghast ; Miller & Young.

P. polygama Walt. f. 2287.
Queens: Hicksville, Hulst ; Jamaica, Brainerd. *Suffolk:* Moriches, Brainerd ; Greenport, Tillinghast ; East Hampton, Mrs. L. D. Pychowska ; Miller & Young.

EUPHORBIACEAE.
ACALYPHA L.
A. Virginica L.. f. 2298.
Kings: Cypress Hills, Hulst. *Queens:* Calverley ; Flushing, Hempstead, N. Hempstead, Oyster Bay, Bisky. *Suffolk:* Greenport, Tillinghast ; Miller & Young.

RICINUS L.
R. communis L. f. 2304.
Common throughout.

EUPHORBIA L.
E. polygonifolia L.. f. 2307.
Queens: Rockaway, Hulst ; Rockaway Park ; Hempstead, Oyster Bay, Bisky. *Suffolk:* Greenport, Tillinghast ; Sag Harbor, E. Hampton, Mrs. L. D. Pychowska ; Miller & Young.

E. nutans Lag. f. 2319.
Kings: Brooklyn, Hulst. *Queens:* Flushing, N. Hempstead, Oyster Bay, Bisky. *Suffolk:* Miller & Young.

E. Ipecacuanhae L. f. 2325.
Queens: Woodhaven, Hulst ; Hempstead, Jamaica, Oyster Bay, Bisky, Valley Stream. *Suffolk:* Miller & Young.

E. Cyparissias L. f. 2337.
Frequent throughout the island.

CALLITRICHACEAE.
CALLITRICHE L.
C. palustris L. f. 2340.
Queens: Calverley, Rudkin.

C. heterophylla Pursh. f. 2341.
  Kings: Forbell's Landing, Hulst. *Queens:* Cold Spring, Hulst; Flushing, Hempstead, Newtown, Oyster Bay, Bisky. *Suffolk:* Miller & Young.

EMPETRACEAE.

COREMA Don.
C. Conradii Torr. f. 2344.
  Queens: Bisky, no definite station recorded. (Specimen?) *Suffolk:* C. W. Eddy, station since lost.

ANACARDIACEAE.

RHUS L.
R. copallina L. f. 2347.
  Common throughout the island.
R. hirta (L.) Sudw. f. 2348.
  Queens: Flushing, Newtown, Bisky.
R. glabra L. f. 2349.
  Common throughout the island.
R. Vernix L. f. 2352.
  Kings: Forbell's Landing, Hulst. *Queens:* Maspeth, Hulst; Throughout, Bisky. *Suffolk:* East Hampton, Mrs. L. D. Pychowska; Miller & Young.
R. radicans L. f. 2353.
  Common throughout the island.

ILICACEAE.

ILEX L.
I. opaca Ait. f. 2356.
  Kings: Prospect Park. *Queens:* Rockaway Beach, Hulst; Flushing, Hempstead, Oyster Bay, Bisky. *Suffolk:* East Hampton, Mrs. L. D. Pychowska; Miller & Young.
I. glabra (L.) Gray. f. 2359.
  Kings: Forbell's Landing, Hulst. *Queens:* Hempstead, Bisky. *Suffolk:* Sag Harbor, Mrs. L. D. Pychowska; Miller & Young.
I. verticillata (L.) Gray. f. 2362.
  Common throughout the island.
I. laevigata (Pursh) Gray. f. 2363.
  Queens: Hempstead, Jamaica, N. Hempstead, Oyster Bay, Bisky. *Suffolk:* Sag Harbor, Mrs. L. D. Pychowska; Miller & Young.

ILICIOIDES Dumont.
I. mucronatum (L.) Britton. f. 2364.
  Queens: Flushing, Newtown, Bisky. *Suffolk:* Greenport, Tillinghast; Miller & Young.

CELASTRACEAE.

EUONYMUS L.
- E. Europaeus L. f. 2368.
  Sparingly escaped.

CELASTRUS L.
- C. scandens L. f. 2370.
  Common throughout the island.

STAPHYLEA L.
- S. trifolia L. f. 2371.
  Kings: Prospect Park, in cultivation.

ACERACEAE.

ACER L.
- A. saccharinum L. f. 2372.
  Common throughout the western end of the island.
- A. rubrum L. f. 2372.
  Common throughout the island.
- A. Saccharum Marsh. f. 2375
  Kings: Hulst, Prospect Park. *Queens:* Richmond Hill, Jamaica. *Suffolk:* Greenport, Tillinghast.
- A. Pennsylvanicum L. f. 2378.
  Queens: Plandome, Eddy.
- A. Negundo L. f. 2380.
  Kings: Prospect Park. *Queens:* Flushing, Bisky.

HIPPOCASTANEACEAE.

AESCULUS L.
- A. Hippocastanum L. f. 2381.
  Frequent throughout.
- A. Pavia L. f. 2385
  Kings: Prospect Park, in cultivation.

BALSAMINACEAE.

IMPATIENS L.
- I. biflora Walt. f. 2388.
  Common throughout the island.

RHAMNACEAE.

RHAMNUS L.
- R. cathartica L. f. 2391.
  Queens: Oyster Bay, Bisky; Glen Cove, Coles. *Suffolk:* Miller & Young.
- R. Frangula L. f. 2395.
  Queens: Flushing, Newtown, Bisky.

CEANOTHUS L.
   C. Americanus L. f. 2396.
      Common throughout the island.

VITACEAE.
VITIS L.
   V. Labrusca L. f. 2398.
      Common throughout the island.
   V. aestivalis Mx. f. 2399.
      Frequent throughout the island.

PARTHENOCISSUS Planch.
   P. quinquefolia (L.) Planch. f. 2410.
      Common throughout the island, especially over the sands at the western end of the island.

TILIACEAE.
TILIA L.
   T. Americana L. f 2411.
      Common throughout the island.
   T. pubescens Ait. f. 2412.
      Moist woods, Long Island, Britton & Brown.

MALVACEAE.
ALTHAEA L.
   A. officinalis L. f. 2414.
      Queens: Flushing, Newtown, N. Hempstead, Oyster Bay, Bisky.
      Suffolk: Miller & Young.

MALVA L.
   M. sylvestris L. f 2415.
      Suffolk: Miller & Young.
   M. rotundifolia L. f. 2416.
      Frequent throughout the island.
   M. moschata L. f. 2418.
      Queens: Cold Spring, Hulst. *Suffolk:* Miller & Young.

ABUTILON Gaertn.
   A. Abutilon (L.) Rusby. f. 2430.
      Frequent throughout the island.

KOSTELETZKYA Presl.
   K. Virginica (L.) A. Gray. f. 2433.
      Queens: Oyster Bay, Mitchell.

HIBISCUS L.
   H. Moscheutos L. f. 2434.
      Frequent throughout the island.

H. Trionum L. f. 2437.
Suffolk: Miller & Young.
H. Syriacus L. f. 2438.
Suffolk: Greenport, Tillinghast; Miller & Young.

HYPERICACEAE.

ASCYRUM L.
A. stans Mx. f. 2442.
Suffolk: Miller & Young.
A. hypericoides L. f. 2443.
Suffolk: Miller & Young.

HYPERICUM L.
H. perforatum L. f. 2454.
Common throughout the island.
H. maculatum Walt. f. 2455.
Long Island, Hulst. *Suffolk:* Greenport, Tillinghast; Miller & Young.
H. mutilum L. f. 2458.
Kings: Calverley, Forbells; Hulst. *Queens.* *Suffolk:* Greenport, Tillinghast; Miller & Young.
H. majus (Gray) Britton. f. 2460.
Suffolk: Miller & Young.
H. Canadense L. f. 2461.
Kings: Cypress Hills, Hulst; Calverley. *Queens.* *Suffolk:* Greenport, Tillinghast; E. Hampton, Mrs. L. D. Pychowska. Miller & Young.

SAROTHRA L.
S. gentianoides L. f. 2463.
Kings: Forbell's Landing, Hulst. *Queens:* Calverley. *Suffolk:* Greenport, Tillinghast; Miller & Young; E. Hampton, Mrs. L. D. Pychowska.

TRIADENUM Raf.
T. Virginicum (L.) Raf. f. 2464.
Kings: Brooklyn. *Queens:* Calverley, Richmond Hill. *Suffolk:* Greenport, Tillinghast; E. Hampton, Mrs. L. D. Pychowska; Miller & Young.

CISTACEAE.

HELIANTHEUM Pers.
H. majus (L.) B.S.P. f. 2470.
Kings: Forbell's Landing, Hulst. *Queens:* Woodhaven, Hulst, Brainerd; Flushing, Hempstead, Newtown, N. Hempstead, Jamaica, Oyster Bay, Bisky. *Suffolk:* Greenport, Tillinghast; Miller & Young.

HUDSONIA L.
  H. ericoides L. f. 2473.
    Kings: Coney Island, Brainerd. *Suffolk:* Southold, Tillinghast; E. Hampton, Mrs. L. D. Pychowska; Miller & Young.
  H. tomentosa Nutt. f. 2474.
    Kings: Coney Island, Brainerd. *Queens:* Rockaway Beach; J. Schrenk; Hempstead, Oyster Bay, Bisky. *Suffolk:* Southold, Tillinghast; E. Hampton, Mrs. L. D. Pychowska; Miller & Young.

LECHEA L.
  L. minor L. f. 2475.
    Kings: Forbell's Landing, Torrey Bot. Club, 1891. *Queens:* Torrey Club, Flushing, Hempstead, Jamaica, Oyster Bay, Bisky; Rockaway. *Suffolk:* E. Hampton, Mrs. L. D. Pychowska; Miller & Young,
  L. racemulosa Mx. f. 2476.
    Queens: Hempstead, Bisky. *Suffolk:* E. Hampton, Mrs. L. D. Pychowska.
  L. maritima Leggett. f. 2478.
    Queens: Hempstead, Oyster Bay, Bisky; Rockaway Beach. *Suffolk:* E. Hampton, Mrs. L. D. Pychowska; Miller & Young.
  L. tenuifolia Mx. f. 2479.
    Kings: Gowanus, Brainerd; Hulst.
  L. Leggettii Britt. & Hall. f. 2480.
    Kings: Forbell's Landing, Hulst. *Queens:* Ridgewood, Rudkin. *Suffolk:* Miller & Young.

VIOLACEAE.

VIOLA L.
  V. palmata L. f. 2484.
    Common throughout the island, more particularly on north side.
  V. sagittata Ait. f. 2490.
    Kings: Greenwood, Graf; Brooklyn. *Queens:* Aqueduct, Forbell's, Hicksville, Hulst; Flushing, Newtown, Bisky. *Suffolk:* Greenport, Tillinghast; Miller & Young.
  V. pedata L. f. 2492.
    Kings: Cedarhurst, Torrey Club. *Queens:* Cypress Hills, Woodhaven, Hulst; Flushing, Hempstead, Oyster Bay, Bisky, *Suffolk:* Greenport, Tillinghast, Miller & Young; East Hampton, Mrs. L. D. Pychowska.
  V. rotundifolia Mx. f. 2494.
    Queens: Long Island City, Hon. A. Brown.

V. blanda Willd. f. 2497.
  Queens: Hulst; Flushing, Oyster Bay, Bisky. *Suffolk:* Miller & Young.
V. primulaefolia L. f. 2499.
  Common throughout the island.
V. lanceolata L. f. 2500.
  Common throughout the island.
V. pubescens Ait. f. 2503.
  Common throughout the island.
V. scabriuscula (T. & G.) Schwein. f. 2504.
  Kings: Ridgewood Heights, Hulst.
V. Labradorica Schrank. f. 2507.
  Queens: Flushing, Oyster Bay, Bisky.
V. tricolor L. f. 2511.
  Throughout island.

CACTACEAE.
OPUNTIA Mill.
  O. Opuntia (L.) Coult. f. 2527.
  Queens: Hickville, Rockaway, Hulst; Oyster Bay, Bisky. *Suffolk:* Greenport,· Tillinghast; Miller & Young; Sag Harbor, Mrs. L. D. Pychowska.

THYMELEACEAE.
DIRCA L.
  D. palustris L. f. 2535.
  Long Island, Hulst.

LYTHRACEAE.
ROTALA L.
  R. ramosier (L.) Koehne. f. 2543.
  Kings: Cypress Hills, Hulst; Ridgewood, Rugers. *Queens:* Newtown, Bisky. *Suffolk:* Miller & Young,
DECODEN J. F. Gmel.
  D. verticillatus (L.) Ell. f. 2544.
  Queens: Jamaica, N. Hempstead, Oyster Bay, Bisky. *Suffolk:* Greenport, Tillinghast; Miller & Young; Sag Harbor, Mrs. L. D. Pychowska.
LYTHRUM L.
  L. Salicaria L. f. 2549.
  Kings: Erie Basin, Hulst. *Queens:* Flushing, Bisky.
PARSONIA P. Br.
  P. petiolata (L.) Rusby. f. 2550.
  Kings: Flatbush, Brainerd.

MELASTOMACEAE.

RHEXIA L.
  R. Virginica L. f. 2552.
    Common throughout the island. Gowanus, Brainerd, 1866.

ONOGRACEAE.

ISNARDIA L.
  I. palustris L. f. 2555.
    Long Island, Hulst. *Queens:* Throughout, Bisky. *Suffolk:* Greenport, Tillinghast; Miller & Young.

LUDWIGIA L.
  L. sphaerocarpa Ell. f. 2557.
    Suffolk: Miller & Young.
  L. alternifolia L. f. 2563.
    Common throughout the island.

CHAMAENERION Adans.
  C. angustifolium (L.) Scop. f. 2566.
    Kings: Forbell's Landing, Hulst. *Queens:* Flushing, Hempstead, Oyster Bay, Bisky. *Suffolk:* Greenport, Tillinghast; Miller & Young.

EPILOBIUM L.
  E. palustre L. f. 2571.
    Queens: Flushing, Bisky. *Suffolk:* Miller & Young.
  E. strictum Muhl. f. 2573.
    Queens: Poggenberg.
  E. coloratum Muhl. f. 2575.
    Common throughout the island.

ONAGRA Adans.
  O. biennis (L.) Scop. f. 2579.
    Common throughout the island.

KNEIFFIA Spach.
  K. Alleni (Britton) Small. f. 2588.
    Suffolk: Miller & Young.
  K. linearis (Michx.) Spach. f. 2590.
    Queens: Glen Cove, Coles. *Suffolk:* E. Hampton, Mrs. L. D. Pychowska; Miller & Young.
  K. pumila (L.) Spach. f. 2591.
    Kings: Prospect Park, Forbell's Landing, Hulst. *Queens:* Flushing, Jamaica, Bisky. *Suffolk:* Miller & Young.
  K. fruticosa (L.) Raimann. f. 2592.
    Queens: Woodhaven, Richmond Hill, Hulst, Hempstead, Jamaica, Oyster Bay, Bisky. *Suffolk:* Greenport, Tillinghast; Miller & Young.

CIRCAEA L.
  C. Lutetiana L. f. 2610.
    Common throughout the island.

HALORAGIDACEAE.

PROSERPINACA L.
  P. palustris L. f. 2615.
    Long Island, Hulst. *Queens:* Flushing, Jamaica, Newtown, Oyster Bay, Bisky. *Suffolk:* Greenport, Tillinghast; Miller & Young.
  P. pectinata Lam. f. 2616.
    Suffolk: Manor, C. H. Peck, New York State Reports, 1889.

MYRIOPHYLLUM L.
  M. tenellum Bigel. f. 2619.
    Queens: Cold Spring, Hulst. *Suffolk:* Miller & Young.
  M. humile (Raf.) Morong. f. 2621.
    Queens: Cold Spring, Hulst; N. Hempstead, Bisky. *Suffolk:* Miller & Young.

ARALIACEAE.

ARALIA L.
  A. spinosa L. f. 2626.
    Kings: Prospect Park. *Queens:* Oyster Bay, Bisky.
  A. racemosa L. f. 2627.
    Queens: Richmond Hill, Hulst; Flushing, Oyster Bay, Bisky.
  A. nudicaulis L. f. 2628.
    Common throughout the island.
  A. hispida Vent. f. 2629.
    Kings: Cypress Hills, Hulst. *Queens:* Flushing, Newtown, Bisky, Richmond Hill.

PANAX L.
  P. trifolium L. f. 2631.
    Kings: Prospect Park. *Queens:* Richmond Hill; Flushing, N. Hempstead, Oyster Bay, Bisky.

UMBELLIFERAE.

DAUCUS L.
  D. Carota L. f. 2632.
    Common throughout the island.

ANGELICA L.
  A. atropurpurea L. f. 2636.
    Kings: Forbell's Landing, Hulst; Prospect Park. *Queens:* Flushing, Oyster Bay, Bisky. *Suffolk:* Miller & Young.

A. villosa (Walt.) B.S.P. f. 2637.
Kings: Cypress Hills, Hulst. Queens: Jamaica, Brainerd; Throughout, Bisky.

OXYPOLIS Raf.
O. rigidus (L.) Britton. f. 2640.
Kings: Cypress Hills, Hulst. Queens: Flushing, Hempstead, Bisky. Suffolk: Miller & Young.

HERACLEUM L.
H. lanatum Mx. f. 2641.
Queens: Flushing, N. Hempstead, Oyster Bay, Bisky. Suffolk: Miller & Young.

PASTINACA L.
P. sativa L. f. 2642.
Common throughout the island.

THASPIUM Nutt.
T. trifoliatum aureum (Nutt.) Britt. f. 2651.
Kings: Gravesend, Brooklyn. Queens: Richmond Hill, Hulst; Flushing, Newtown, Oyster Bay, Bisky. Suffolk: Miller & Young.

AETHUSA L.
A. Cynapium L. f. 2656.
Queens: Flushing, Bisky.

SANICULA L.
S. Marylandica L. f. 2664.
Common throughout the island.
S. Canadensis L. f. 2666.
Common throughout the island.

PIMPINELLA L.
P. integerrima (L.) Gray. f. 2670.
Kings: Forbell's Landing, Hulst. Queens: Newtown, Bisky.

CHAEROPHYLLUM L.
C. procumbens (L.) Crantz. f. 2678.
Queens: Richmond Hill, Hulst.

WASHINGTONIA Raf.
W. Claytoni (Mx.) Britton. f. 2680.
Long Island, Hulst. Queens: Flushing, Oyster Bay, Bisky.
W. longistylis (Torr.) Britton. f. 2681.
Common throughout northern shore of island.

CONIUM L.
C. maculatum L. f. 2684.
Kings: Forbell's Landing, fide Chas. H. Hall; com. by Dr. Hulst.

SIUM L.
  S. cicutaefolium Gmel. f. 2685.
    Kings: Calverley ; Forbell's Landing, Hulst. *Queens:* Flushing, Jamaica, Bisky. *Suffolk:* Miller & Young.

ZIZIA Koch.
  Z. aurea (L.) Koch. f. 2690.
    Long Island, Hulst.

CARUM L.
  C. Carui L. f. 2693.
    Suffolk: Miller & Young.

CICUTA L.
  C. maculata L. f. 2694.
    Common throughout the island.
  C. bulbifera L. f. 2695.
    Suffolk: Miller & Young.

DERINGA Adans.
  D. Canadensis (L.) Kuntz. f. 2696.
    Kings: Prospect Park. *Queens:* Flushing, Oyster Bay, Bisky.

PTILIMNIUM Raf.
  P. capillaceum (Mx.) Hollick. f. 2699.
    Kings: New Lots, Brainerd. *Queens:* Flushing, Hempstead, N. Hempstead, Oyster Bay, Bisky. *Suffolk:* East Hampton, Mrs. L. D. Pychowska ; Miller & Young.

HYDROCOTYLE L.
  H. umbellata L. f. 2703.
    Suffolk: Miller & Young.
  H. Americana L. f. 2706.
    Kings: Forbell's Landing, Hulst. *Queens:* Cold Spring, Hulst : Calverley ; Flushing, Oyster Bay, Bisky.

CORNACEAE.

CORNUS L.
  C. Canadensis L. f. 2710.
    Long Island, Hulst. *Queens:* Newtown, Bisky ; Dutch Kills, M. Rugers. *Suffolk:* Miller & Young.
  C. florida L. f. 2712.
    Common throughout the island.
  C. Amomum Mill. f. 2714.
    Queens: Maspeth, Hulst ; Flushing, Oyster Bay, Bisky. *Suffolk:* Miller & Young.
  C. stolonifera Mx. f. 2717.
    Kings: Prospect Park. *Queens:* Winfield, Hulst ; Flushing, Bisky. *Suffolk:* Miller & Young.

C. candidissima Marsh. f. 2718.
Kings: Forbell's Landing, Hulst. *Queens:* Flushing, Bisky.

C. alternifolia L. f. f. 2720.
Queens: Richmond Hill, Hulst; Flushing, Newtown, N. Hempstead, Oyster Bay, Bisky. *Suffolk:* Miller & Young.

NYSSA L.

N. sylvatica Marsh. f. 2721
Queens: throughout, Bisky. *Suffolk:* Greenport, Tillinghast; Miller & Young; E. Hampton, Mrs. L. D. Pychowska.

GAMOPETALAE.

CLETHRACEAE.

CLETHRA L.

C. alnifolia L. f. 2724.
Frequent throughout the island.

PYROLACEAE.

PYROLA L.

P. rotundifolia L. f. 2726.
Rare throughout the island.

P. chlorantha Sw. f. 2727.
Suffolk: Miller & Young.

P. elliptica Nutt. f. 2728.
Frequent throughout the island.

P. secunda L. f. 2733.
Queens: Jamaica, Bisky.

CHIMAPHILA Pursh.

C. maculata (L.) Pursh. f. 2735.
Frequent throughout the island.

C. umbellata (L.) Nutt. f. 2736.
Frequent throughout the island.

MONOTROPACEAE.

MONOTROPA L.

M. uniflora L. f. 2739.
Kings: Calverley; Hulst. *Queens:* Woodhaven, Richmond Hill; Flushing, Hempstead, Oyster Bay, Bisky. *Suffolk:* Miller & Young.

HYPOPITYS Adans.

H. Hypopitys (L.) Small. f. 2740.
Kings: Calverley. *Queens:* Calverley; Cold Spring, Hulst: Flushing, Hempstead, Oyster Bay, Bisky. *Suffolk:* Miller & Young.

ERICACEAE.

AZALEA L.
- **A. nudiflora** L. f. 2743.
  Frequent throughout the island.
- **A. viscosa** L. f. 2747.
  Common throughout the island.
- **A. viscosa hispida** (Pursh) Britton.
  Montauk Point, Britton & Brown.
- **A. viscosa nitida** (Pursh) Britton.
  Long Island, Hulst. *Queens:* Bisky.
- **A. viscosa glauca** Mx.
  Queens: Flushing, Hempstead, Bisky.

RHODORA L.
- **R. Canadense** L. f. 2748.
  Queens: Flushing, Torrey.

RHODODENDRON L.
- **R. maximum** L. f. 2750.
  Extensively cultivated. *Suffolk:* Miller & Young.

KALMIA L.
- **K. angustifolia** L. f. 2756.
  Queens: Richmond Hill, Hulst; Hempstead, Bisky. *Suffolk:* Greenport, Tillinghast; Miller & Young.
- **K. latifolia** L. f. 2757.
  Kings: Forbell's Landing, Hulst. *Queens:* Cold Spring, Hulst; Flushing, Oyster Bay, Bisky. *Suffolk:* Greenport, Tillinghast; Miller & Young.

LEUCOTHOË D. Don.
- **L. racemosa** (L.) Gray. f. 2766.
  Kings: Forbell's Landing, Hulst. *Queens:* Flushing, Newtown, Bisky. *Suffolk:* Greenport, Tillinghast; Miller & Young.

PIERIS D. Don.
- **P. Mariana** (L.) Benth. & Hook. f. 2770.
  Kings: Cypress Hills, Hulst; New Lots. *Queens:* Calverley, Hicksville, Hulst; Hempstead, Jamaica, Bisky. *Suffolk:* E. Hampton, Mrs. L. D. Pychowska; Miller & Young.

XOLISMA Raf.
- **X. ligustrina** (L.) Britton. f. 2771.
  Queens: Woodhaven, Hulst; throughout, Bisky. *Suffolk:* Miller & Young; E. Hampton, Sag Harbor, Mrs. L. D. Pychowska.

CHAMAEDAPHNE Moench.
- **C. calyculata** (L.) Moench. f. 2772.
  Queens: Newtown, Prof. Schrenk.

EPIGAEA L.
  E. repens L. f. 2774.
    Frequent throughout the island.

GAULTHERIA L.
  G. procumbens L. f. 2775.
    Kings: Calverley; Cypress Hills, Hulst. *Queens:* Flushing, Hempstead, Oyster Bay, Bisky. *Suffolk:* Miller & Young.

ARCTOSTAPHYLOS Adans.
  A. Uva-Ursi (L.) Spreng. f. 2776.
    Queens: Cold Spring, Hulst; Hempstead, Oyster Bay, Bisky. *Suffolk:* E. Hampton, Mrs. L. D. Pychowska; Miller & Young.

VACCINIACEAE.

GAYLUSSACIA H.B.K.
  G. frondosa (L.) T. & G. f. 2779.
    Kings: Prospect Park. *Queens:* Woodhaven, Hulst; Hempstead, Jamaica, Bisky. *Suffolk:* Miller & Young.
  G. resinosa (Ait.) T. & G. f. 2780.
    Kings: Prospect Park. *Queens:* throughout, Bisky. *Suffolk:* Wading River, Brainerd; Miller & Young.
  G. dumosa (Andr.) T. & G. f. 2781.
    Suffolk: Miller & Young.

VACCINIUM L.
  V. corymbosum L. f. 2788.
    Common throughout the island.
  V. atrococcum (Gray) Heller. f. 2789
    Queens: Bisky. *Suffolk:* Wading River, Brainerd.
  V. Canadense Richards. f. 2790.
    Queens: Flushing, Bisky.
  V. Pennsylvanicum Lam. f. 2791.
    Queens: Woodhaven, Hulst; Glen Cove, Coles. *Suffolk:* Wading River, Hulst, Miller & Young.
  V. vacillans Soland. f. 2793.
    Queens: throughout. Bisky. *Suffolk:* Miller & Young.
  V. stamineum L. f. 2796.
    Kings: Forbell's Landing, Hulst.

CHIOGENES Salisb.
  C. hispidula (L.) T. & G. f. 2798.
    Long Island, *fide* Dr. Chas. H. Hall.

OXYCOCCUS Hill.
  O. macrocarpus (Ait.) Pers. f. 2800.
    Kings: Cedarbush, Hulst. *Queens:* Hempstead, Oyster Bay, Bisky. *Suffolk:* E. Hampton, Mrs. L. D. Pychowska; Miller & Young.

PRIMULACEAE.

HOTTONIA L.
    H. inflata Ell. f. 2809.
        Queens: Calverley. *Suffolk:* Miller & Young.

SAMOLUS L.
    S. floribundus H.B.K. f. 2810.
        Kings: Hulst. *Queens:* Flushing, Hempstead, Oyster Bay, Bisky. *Suffolk:* Miller & Young.

LYSIMACHIA L.
    L. vulgaris L. f. 2811.
        Queens: Newtown, Rudkin.
    L. quadrifolia L. f. 2813.
        Common throughout the island.
    L. terrestris (L.) B.S.P. f. 2814.
        Frequent throughout the island.
    L. Nummularia L. f. 2815.
        Frequent throughout the island.

STEIRONEMA Raf.
    S. ciliatum (L.) Raf. f. 2816.
        Kings: New Lots, Brainerd; Hulst. *Queens:* Flushing, Newtown, Bisky.
    S. lanceolatum (Walt.) Gray. f. 2819.
        Queens: Flushing, Oyster Bay, Bisky. *Suffolk:* Miller & Young.

NAUMBURGIA Moench.
    N. thyrsiflora (L.) Duby f. 2821.
        Kings: Forbell's Landing, Hulst; Vandewern Creek, Rudkin. *Queens:* Glen Cove, Coles; Woodside, Brown.

TRIENTALIS L.
    T. Americana Pursh. f. 2822.
        Common throughout the island.

ANAGALLIS L.
    A. arvensis L. f. 2824.
        Frequent throughout the island.

DODECATHEON L.
    D. Meadia L. f. 2826.
        Queens: Glen Cove, Coles.

PLUMBAGINACEAE.

LIMONIUM Adans.
    L. Carolinianum (Walt.) Britton. f. 2827.
        Common in the sea marshes throughout the island.

EBENACEAE.

DIOSPYROS L.
   D. Virginiana L. f. 2831.
      Kings: New Lots, Hulst; Flatbush. *Queens:* Oyster Bay, Bisky.

OLEACEAE.

SYRINGA L.
   S. vulgaris L. f. 2837.
      Common throughout the island.

FRAXINUS L.
   F. Americana L. f. 2838.
      Kings: Brooklyn, Hulst; *Queens:* Flushing, Oyster Bay, Bisky.
   F. Pennsylvanica Marsh. f. 2840.
      Queens: Maspeth, Hulst.
   F. nigra Marsh. f. 2843.
      Queens: Glen Cove, Coles.

CHIONANTHUS L.
   C. Virginica L. f. 2845.
      Kings: Prospect Park (cult.)?

LIGUSTRUM L.
   L. vulgare L. f. 2846.
      Frequent throughout the island.

GENTIANACEAE.

ERYTHRAEA Neck.
   E. pulchella (Sw.) Fries. f. 2853.
      Queens: Flushing, W. Cooper, 1832, in herb. Torrey.

SABBATIA Adans.
   S. angularis (L.) Pursh. f. 2858.
      Queens: Richmond Hill, Hulst; Glendale, Ruger.
   S. stellaris Pursh. f. 2861.
      Frequent along the salt meadows.
   S. campanulata (L.) Torr. f. 2863.
      Kings: Forbell's Landing, Hulst.
   S. dodecandra (L.) B.S.P. f. 2864.
      Kings: Forbell's Landing, Hulst; New Lots, J. F. Poggenburg.
      Queens: Richmond Hill?

GENTIANA L.
   G. crinita Froel. f. 2867.
      Kings: Forbell's Landing, Hulst. *Queens:* Richmond Hill; Aqueduct, Hulst; Flushing, Newtown, Oyster Bay, Bisky. *Suffolk:* Miller & Young; E. Hampton, Mrs. L. D. Pychowska.

G. Saponaria L. f. 2875.
   Queens: Richmond Hill.
G. Andrewsii Griseb. f. 2876.
   Kings: Forbell's Landing, Hulst. *Queens*: Flushing, Oyster Bay, Bisky.

BARTONIA Muhl.
B. Virginica (L.) B.S.P. f. 2887.
   Queens: Hicksville, Hulst; Mrs. E. G. Britton; J. Schrenk.

MENYANTHACEAE.

MENYANTHES L.
M. trifoliata L. f. 2889.
   Kings: Forbell's Landing, Hulst; New Lots, Hulst, Brainerd; Calverley. *Queens*: Woodhaven, Torrey Club; Flushing, Newtown, Bisky. *Suffolk*: Miller & Young.

LIMNANTHEMUM S. G. Gmelin.
L. lacunosum (Vent) Griseb. f. 2890.
   Suffolk: Miller & Young.

APOCYNACEAE.

VINCA L.
V. minor L. f. 2894.
   Kings: Common throughout the county, escaped from gardens, cemeteries, etc.

APOCYNUM L.
A. androsaemifolium L. f. 2895.
   Queens: Woodhaven, Hulst; Flushing, Oyster Bay, Bisky. *Suffolk*: Greenport, Tillinghast, Miller & Young.
A. cannabinum L. f. 2896.
   Kings: Cypress Hills, Hulst. *Queens*: Flushing, Bisky. *Suffolk*: Greenport, Tillinghast, Miller & Young.

ASCLEPIADACEAE.

ASCLEPIAS L.
A. tuberosa L. f. 2900.
   Frequent throughout the island.
A. rubra L. f. 2903.
   Queens: Centreville, Rugers; Glen Cove, Coles.
A. purpurascens L. f. 2904.
   Kings: Forbell's Landing, Hulst. *Queens*: Flushing, Oyster Bay, Bisky. *Suffolk*: Greenport, Tillinghast; Miller & Young.
A. incarnata L. f. 2905.
   Frequent throughout the island.

A. obtusifolia Mx. f. 2909.
   Queens: Woodhaven, Hulst; Hempstead, Jamaica, Newtown, Oyster Bay, Bisky. *Suffolk:* Greenport, Tillinghast; Miller & Young; East Hampton, Mrs. L. D. Pychowska.

A. exaltata (L.) Muhl. f. 2911.
   Kings: Forbell's Landing, Hulst. *Queens:* Leggett. *Suffolk:* Miller & Young.

A. variegata L. f. 2912.
   Kings: Hulst. *Queens:* Jamaica, Brainerd; N. Hempstead, W. H. Leggett; Glen Cove, Coles. *Suffolk:* Miller & Young; Sag Harbor, Mrs. L. D. Pychowska.

A. quadrifolia Jacq. f. 2913.
   Queens: Jamaica, Brainerd; Richmond Hill; Flushing, Newtown, Oyster Bay, Bisky. *Suffolk:* Sag Harbor, Mrs. L. D. Pychowska.

A. Syriaca L. f. 2914.
   Common throughout the island.

A. verticillata L. f. 2920.
   Queens: Hicksville, Hulst; Flushing, Hempstead, Newtown, N. Hempstead, Bisky. *Suffolk:* Miller & Young; Sag Harbor, Mrs. L. D. Pychowska.

ACERATES Ell

A. viridiflora (Raf.) Eaton. f. 2924.
   Queens: Aqueduct, Hulst; Centreville, Rugers.

CYNANCHUM L.

C. nigrum (L.) Pers. f. 2930.
   Kings: Cypress Hills, Hulst. *Queens:* Flushing, Bisky.

CONVOLVULACEAE.

IPOMOEA L.

I. purpurea (L.) Roth. f. 2949.
   Frequent throughout the island.

CONVOLVULUS L.

C. sepium L. f. 2951.
   Kings: Forbell's Landing, Hulst. *Queens:* Flushing, Oyster bay, Bisky; Flushing, Brainerd. *Suffolk:* Miller & Young; E. Hampton, Mrs. L. D. Pychowska.

C. repens L. f. 2952.
   Queens: Mrs. E. G. Britton.

C. spithamaeus L. f. 2953.
   Queens: Woodhaven, Ruger.

C. arvensis L. f. 2954.
   Frequent throughout the island.

CUSCUTACEAE.

CUSCUTA L.
  C. arvensis Beyrich. f. 2958.
    Queens: Hempstead, Bisky.
  C. Gronovii Willd. f. 2963.
    Frequent throughout the island.
  C. compacta Juss. f. 2966.
    Kings: Gowanus, Brainerd.

POLEMONIACEAE.

PHLOX L.
  P. paniculata L. f. 2968.
    Frequent throughout the island.
  P. subulata Mx. f. 2979.
    Kings: Cypress Hills, Hulst.

BORAGINACEAE.

CYNOGLOSSUM L.
  C. officinale L. f. 3019.
    Queens: Cold Spring, Hulst; Flushing, Oyster Bay, Bisky.
  C. Virginicum L. f. 3020.
    Queens: Jamaica, Brainerd; Maspeth, Hulst.
LAPPULA Moench.
  L. Lappula (L.) Karst. f. 3021.
    Long Island, Brainerd.
  L. Virginianum (L.) Greene. f. 3023.
    Kings: Hulst. *Queens:* Flushing, Bisky.
MYOSOTIS L.
  M. palustris (L.) Lam. f. 3038.
    Occasionally escaped from cultivation throughout the island.
    Kings: Calverley; Forbell's Landing, Hulst. *Queens:* Flushing, N. Hempstead, Oyster Bay, Bisky.
  M. laxa Lehm. f. 3039.
    Kings: Forbell's Landing, Hulst. *Suffolk:* Miller & Young.
  M. Virginica (L.) B.S.P. f. 3042.
    Kings: Greenwood, Brainerd. *Queens:* Flushing, Newtown, Bisky.
LITHOSPERMUM L.
  L. arvense L. f. 3043.
    Queens: Woodhaven, Hulst; Flushing, Bisky. *Suffolk:* Greenport, Tillinghast; Miller & Young.
  L. officinale L. f. 3044.
    Kings: Gowanus, Brainerd.

ONOSMODIUM Mx.
   O. Virginianum (L.) DC.   f. 3052.
      Queens: Jamaica, Leggett.   *Suffolk:* Miller & Young.
SYMPHYTUM L.
   S. officinale L.   f. 3053.
      Queens: Winfield, Hulst; Flushing, Bisky.   *Suffolk:* Miller & Young.
LYCOPSIS L.
   L. arvensis L.   f. 3055.
      Suffolk: Miller & Young.
ECHIUM L.
   E. vulgare L.   f. 3056.
      Frequent throughout the island.

VERBENACEAE.

VERBENA L.
   V. officinalis L.   f. 3057.
      Kings: Hulst.   *Suffolk:* Miller & Young.
   V. urtricifolia L.   f. 3058.
      Kings: Flatbush, C. F. Neuhaus.   *Queens:* Woodhaven, Hulst; Flushing, N. Hempstead, Oyster Bay, Bisky.   *Suffolk:* Greenport, Tillinghast; Miller & Young.
   V. hastata L.   f. 3059.
      Frequent throughout the island.
   V. angustifolia Mx.   f. 3060.
      Kings: Calverley.   *Queens:* Jamaica, J. Hall.

LABIATEAE.

TEUCRIUM L.
   T. Canadense L.   f. 3070.
      Frequent throughout the island.
ISANTHUS Mx.
   I. brachiatus (L.) B.S.P.   f. 3072.
      Queens: Woodhaven, Hulst.
TRICHOSTEMA L.
   T. dichotomum L.   f. 3073.
      Queens: Woodhaven, Hulst; Richmond Hill, Miss M. F. Long; throughout, Bisky.   *Suffolk:* Greenport, Tillinghast; Miller & Young; E. Hampton, Mrs. L. D. Pychowska.
   T. lineare Nutt.   f. 3074.
      Long Island, Britton & Brown.
SCUTELLARIA L.
   S. lateriflora L.   f. 3075.
      Frequent throughout the island.

S. cordifolia Muhl.   f. 3078.
    Kings: Hulst.
S. pilosa Mx.   f. 3079.
    Queens: Woodside, Schrenk.
S. integrifolia L.   f. 3080.
    Suffolk: Miller & Young.
S. galericulata L.   f. 3087.
    Frequent throughout the island.

MARRUBIUM L.
M. vulgare L.   f. 3089.
    Queens: Aqueduct, Hulst ; Flushing, Oyster Bay, Bisky.   *Suffolk:* Miller & Young.

AGASTACHE Clayt.
A. nepetoides (L.) Kuntze.   f. 3090.
    Kings: Brainerd.   *Queens:* Maspeth, Hulst ; Flushing, Newtown, Oyster Bay, Bisky.
A. scrophulariaefolia (Willd.) Kuntze.   f. 3091.
    Queens: Mrs. E. G. Britton.   *Suffolk:* Miller & Young.

NEPETA L.
N. Cataria L.   f. 3094.
    Common throughout the island.

GLECOMA L.
G. hederacea L.   f. 3095.
    Frequent throughout the island.

PRUNELLA L.
P. vulgaris L.   f. 3098.
    Common throughout the island.

PHYSOSTEGIA Benth.
P. Virginiana (L.) Benth.   f. 3100.
    Queens: Flushing, Allen.

GALEOPSIS L.
G. Tetrahit L.   f. 3107.
    Queens: Oyster Bay, Bisky.   *Suffolk:* East Hampton, Mrs. L. D. Pychowska.

LEONURUS L.
L. Cardiaca L.   f. 3108.
    Common throughout the island.

LAMIUM L.
L. amplexicaule L.   f. 3111.
    Kings: Brooklyn, Hulst.   *Queens:* Flushing, Newtown, Bisky.   *Suffolk:* Greenport, Tillinghast ; Miller & Young.
L. purpureum L.   f. 3112.
    Queens: Flushing, Bisky.

STACHYS L.
- **S. hyssopifolia** Mx. f. 3116.
  Kings: Brainerd. *Queens:* Hicksville, Hulst; Flushing, Jamaica, Bisky. *Suffolk:* Miller & Young; Sag Harbor, Mrs. L. D. Pychowska.
- **S. palustris** L. f. 3119.
  Queens: Jamaica, Rudkin.

MONARDA L.
- **M. didyma** L. f. 3131.
  Queens: Cold Spring, Hulst.
- **M. fistulosa** L. f. 3133.
  Queens: Ridgewood, Bisky. *Suffolk:* Miller & Young.

HEDEOMA Pers.
- **H. pulegioides** (L.) Pers. f. 3141.
  Frequent throughout the island.

MELISSA L.
- **M. officinalis** L. f. 3144.
  Queens: N. Hempstead, Oyster Bay, Bisky. *Suffolk:* Miller & Young.

KOELLIA Moench.
- **K. flexuosa** (Walt.) MacM. f. 3154.
  Queens: Hicksville, Hulst; Flushing, Oyster Bay, Bisky. *Suffolk:* Miller & Young.
- **K. Virginiana** (L.) MacM. f. 3155.
  Kings: Forbell's Landing, Hulst; New Lots. *Queens:* Jamaica, Oyster Bay, Bisky. *Suffolk:* E. Hampton, Mrs. L. D. Pychowska.
- **K. incana** (L.) Kuntze. f. 3161.
  Queens: Woodside, Schrenk. *Suffolk:* Miller & Young.
- **K. mutica** (Mx.) Britton. f 3164.
  Queens: Woodside, Schrenk. *Suffolk:* E. Hampton, Mrs. L. D. Pychowska; Miller & Young.

CUNILA L.
- **C. origanoides** (L.) Britton. f. 3167.
  Queens: Glen Cove, Bisky. *Suffolk:* Miller & Young.

LYCOPUS L.
- **L. Virginicus** L. f. 3168.
  Frequent throughout the island.
- **L. sessilifolius** Gray. f. 3169.
  Kings: Hulst. *Suffolk:* Riverhead, C. H. Peck, 1889.
- **L. rubellus** Moench. f. 3170.
  Kings: Hulst. *Suffolk:* Miller & Young.

L. Americanus Muhl. f. 3171.
 Queens: Flushing, N. Hempstead, Oyster Bay, Bisky. Suffolk. Greenport, Tillinghast; Miller & Young.
L. Europaeus L. f. 3173.
 Kings: Hulst. Suffolk: Miller & Young.

MENTHA L.
M. spicata L. f. 3174.
 Frequent throughout the island.
M. piperita L. f. 3175.
 Kings: Forbell's Landing, Hulst. Queens: Flushing, N. Hempstead, Oyster Bay, Bisky. Suffolk: Miller & Young.
M. rotundifolia (L.) Huds. f. 3178.
 Suffolk: Miller & Young.
M. crispa L. f. 3181.
 Suffolk: Miller & Young.
M. arvensis L. f. 3182.
 Queens: Flushing, Bisky.
M. sativa L. f. 3184.
 Queens: Oyster Bay, Bisky.
M. Canadensis L. f. 3185.
 Kings: Cypress Hills, Hulst. Queens: Flushing, Bisky. Suffolk: Miller & Young.

COLLINSONIA L.
C. Canadensis L. f. 3186.
 Frequent in moist woods throughout the island.

SOLANACEAE.

PHYSALODES Boehm.
P. Physaloides (L.) Britton. f. 3189.
 Queens: Woodhaven, Ruger.

PHYSALIS L.
P. pubescens L. f. 3190.
 Queens: Richmond Hill, Hulst; Flushing, Newtown, Oyster Bay, Bisky. Suffolk: Miller & Young.
P. Philadelphica Lam. f. 3197.
 Queens: Flushing, Allen.
P. Virginiana Mill. f. 3202.
 Kings: Hulst. Queens: Rockaway Beach, Leggett; Oyster Bay, Bisky. Suffolk: East Hampton, Mrs. L. D. Pychowska.
P. Peruviana L.
 Suffolk: Manor, C. H. Peck, New York State Reports, 1889. Escaped.

SOLANUM L..
  S. nigrum L. f. 3211.
    Queens: Rockaway Beach, Hulst ; Flushing, Oyster Bay, Bisky.
    Suffolk: Miller & Young.
  S. Carolinense L. f. 3213.
    Kings: Prospect Park, Eccles ; Hulst ; J. *Queens:* Flushing, Bisky.
  S. rostratum Dunal. f. 3216.
    Kings: Hulst.
  S. Dulcamara L. f. 3218.
    Common throughout the island.

LYCOPERSICON Mill.
  L. Lycopersicon (L.) Karst. f. 3219.
    Commonly escaped from cultivation.

LYCIUM L.
  L. vulgare (Ait.) Dun. f. 3220.
    Frequent throughout the island.

HYOSCYAMUS L.
  H. niger L. f. 3221.
    Kings: Prospect Park, Cult. *Queens:* Glen Cove, Coles.

DATURA L.
  D. Stramonium L. f. 3222.
    Common throughout the island.
  D. Tatula L. f. 3223.
    Frequent throughout the island.

PETUNIA Juss.
  P. axillaris (Lam.) B.S.P. f. 3227.
    Kings: Cultivated.

SCROPHULARIACEAE.

VERBASCUM L.
  V. Thapsus L. f. 3229.
    Common throughout the island.
  V. Blattaria L. f. 3232.
    Frequent throughout the island.

LINARIA Juss.
  L. Linaria (L.) Karst. f. 3236.
    Common throughout the island.
  L. genistaefolia (L.) Miller. f. 3237.
    Queens: Calverley.
  L. Canadensis (L.) Dumont. f. 3238.
    Frequent throughout the island.

SCROPHULARIA L.
  S. Marylandica L. f. 3242.
    Kings: Hulst. *Queens:* Flushing, Hempstead, Newtown, Oyster Bay, Bisky. *Suffolk:* Greenport, Tillinghast; Miller & Young.

CHELONE L.
  C. glabra L. f. 3244.
    Frequent throughout the island.

PENTSTEMON Soland.
  P. hirsutus (L.) Willd. f. 3247.
    Kings: Hulst.
  P. Pentstemon (L.) Britton. f. 3252.
    Kings: Prospect Park. *Queens:* Glen Cove.

PAULOWNIA Sieb. & Zucc.
  P. tomentosa (Thunb.) Baill. f. 3264.
    Kings: Throughout the city of Brooklyn, cultivated.

MIMULUS L.
  M. ringens L. f. 3265.
    Frequent throughout the island.
  M moschatus Dougl. f. 3269.
    Queens: Oyster Bay, Locust Valley, Bisky.

GRATIOLA L.
  G. Virginiana L. f. 3275.
    Kings: Cypress Hills, Hulst. *Queens:* Flushing, Newtown, Bisky.
  G. aurea Muhl. f. 3277.
    Queens: Cold Spring, Hulst; Hempstead, Jamaica, N. Hempstead, Oyster Bay, Bisky. *Suffolk:* Miller & Young; Napeague Beach, Mrs. L. D. Pychowska.

ILYSANTHES Raf.
  I. gratioloides (L.) Benth. f. 3280.
    Frequent throughout the island.

VERONICA L.
  V. Anagallis-aquatica L. f. 3287.
    Queens: Flushing, Bisky.
  V. Americana Schwein. f. 3288.
    Kings: New Lots, Brainerd; Hulst. *Suffolk:* Miller & Young.
  V. scutellata L. f. 3289.
    Kings: Cypress Hills, Hulst.
  V. officinalis L. f. 3290.
    Common throughout the island.
  V. serpyllifolia L. f. 3293.
    Frequent throughout the island.

V. peregrina L.   f. 3294.
Frequent throughout the island.
V. arvensis L.   f. 3295.
Frequent throughout the island.
V. agrestis L.   f. 3296.
Queens: Glen Cove, Brainerd.
V. Byzantina (Sibth & Smith) B.S.P.   f. 3297.
Queens: Flushing, Bisky.

LEPTANDRA Nutt.
L. Virginica (L.) Nutt.   f. 3299.
Frequent throughout the island.

DIGITALIS L.
D. purpurea L.   f. 3300.
Sparingly escaped from cultivation.

DASYSTOMA Raf.
D. Pedicularia (L.) Benth.   f. 3303.
Queens: Hempstead, Oyster Bay, Bisky. *Suffolk:* Greenport, Tillinghast; Miller & Young; East Hampton, Mrs. L. D. Pychowska.
D. flava (L.) Wood.   f. 3304.
Kings: Cypress Hills, Hulst; Forbell's Landing; *Queens:* Flushing, Oyster Bay, Bisky. *Suffolk:* Greenport, Tillinghast; E. Hampton, Mrs. L. D. Pychowska; Miller & Young.
D. Virginica (L.) Britton.   f. 3307.
Queens: Woodhaven, Hulst; Newtown, N. Hempstead, Bisky. *Suffolk:* Greenport, Tillinghast; Miller & Young.

GERARDIA L.
G. purpurea L.   f. 3310.
Frequent throughout the island.
G. maritima L.   f. 3312.
Frequent along the shores of the island.
G. tenuifolia Vahl.   f. 3313.
Frequent throughout the island.

CASTILLEJA Mutis.
C. coccinea (L.) Spreng.   f. 3318.
Queens: Newtown, Oyster Bay, Bisky.

PEDICULARIS L.
P. Canadensis L.   f. 3335.
Common throughout the island.

MELAMPYRUM L.
M. lineare Lam.   f. 3340.
Frequent throughout the island.

LENTIBULARIACEAE.
UTRICULARIA L.
  U. cornuta Mx. f. 3342.
    Queens: Centreville, Rugers. *Suffolk:* Miller & Young.
  U. resupinata B. D. Greene. f. 3344.
    Suffolk: Miller & Young.
  U. inflata Walt. f. 3347.
    Suffolk: Miller & Young.
  U. purpurea Walt. f. 3348.
    Queens: Ridgwood Cemetery, Bisky. *Suffolk:* Miller & Young; Fore and Aft Pond, Sag Harbor, Mrs. L. D. Pychowska.
  U. vulgaris L. f. 3349.
    Kings: Brooklyn, Brainerd, Hulst. *Queens:* Jamaica, Brainerd; Hempstead, Newtown, Bisky; Glen Cove. *Suffolk:* Miller & Young.
  U. intermedia Hayne. f. 3351.
    Queens: Jamaica, Brainerd. *Suffolk:* Miller & Young.
  U. fibrosa Walt. f. 3352.
    Suffolk: Miller & Young.
  U. gibba L. f. 3354.
    Queens: Woodhaven, Rugers. *Suffolk:* Sag Harbor, Mrs. L. D. Pychowska; Miller & Young.
  U. cana Gray. ?
    Suffolk: Miller & Young.

OROBANCHACEAE.
THALESIA Raf.
  T. uniflora (L.) Britton. f. 3358.
    Kings: Prospect Park; Newtown, Hulst; Greenwood, Brainerd. *Queens:* Richmond Hill; Flushing, Newtown, Huntington, Miss J. E. Rogers; Oyster Bay, Bisky. *Suffolk:* Miller & Young.
LEPTAMNIUM Raf.
  L. Virginianum (L.) Raf. f. 3364.
    Frequent on the island.

BIGNONIACEAE.
TECOMA Juss.
  T. radicans (L.) DC. f. 3366.
    Kings: Prospect Park. *Queens:* Cold Spring, Hulst.
CATALPA Scop.
  C. Catalpa (L.) Karst. f. 3367.
    Kings: Brooklyn, common. *Queens:* Cold Spring, Hulst; Maspeth, Ruger.

MARTYNIACEAE.

MARTYNIA L.
  M. Louisiana Mill. f. 3369.
    Kings: Hulst. *Suffolk:* Miller & Young.

PHRYMACEAE.

PHRYMA L.
  P. Leptostachya L. f. 3377.
    Frequent throughout the island.

PLANTAGINACEAE.

PLANTAGO L.
  P. major L. f. 3378.
    Common throughout the island.
  P. Rugelii Dec. f. 3379.
    Kings: Brooklyn, Hulst.
  P. lanceolata L. f. 3380.
    Common throughout the island.
  P. maritima L. f. 3385.
    Frequent along the coast.
  P. aristata Mx. f. 3387.
    Kings: New Lots, Hulst.   *Queens:* Cold Spring, Hulst.
  P. Virginica L. f. 3388.
    Suffolk: Miller & Young.
  P. elongata Pursh. f. 3389.
    Queens: Newtown, Bisky.   *Suffolk:* Greenport, Tillinghast; Miller & Young.

RUBIACEAE.

HOUSTONIA L.
  H. coerulea L. f. 3393.
    Kings: Cypress Hills, Hulst.   *Queens:* Richmond Hill.
  H. purpurea L. f. 3397.
    Queens: Spring, Rudkin.
  H. longifolia Gaertn. f. 3399.
    Queens: Hempstead, Jamaica, N. Hempstead, Oyster Bay, Bisky.
    Suffolk: Miller & Young.

OLDENLANDIA L.
  O. uniflora L. f. 3402.
    Queens: Hempstead, Bisky; Rockaway.   *Suffolk:* Miller & Young.

CEPHALANTHUS L.
  C. occidentalis L. f. 3403.
    Common throughout the island.

MITCHELLA L.
  M. repens L. f. 3404.
    Common throughout the island.

DIODIA L.
  D. teres Walt. f. 3406.
    Kings: Forbell's Landing, Hulst. *Queens:* Woodhaven, Hulst; Jamaica, Bisky. *Suffolk:* Manor, C. H. Peck, 1890.

GALIUM L.
  G. verum L. f. 3408.
    Queens: Glen Cove, Coles.
  G. Aparine L. f. 3412.
    Kings: Gowanus, Brainerd. *Queens:* Aqueduct, Hulst; N. Hempstead, Oyster Bay, Bisky. *Suffolk:* Greenport, Tillinghast; Miller & Young.
  G. pilosum Ait. f. 3415.
    Kings: New Lots, Brainerd. *Queens:* Newtown, N. Hempstead, Bisky. *Suffolk:* Miller & Young; E. Hampton, Mrs. L. D. Pychowska.
  G. circaezans Michx. f. 3417.
    Kings: Cypress Hills, Hulst; New Lots, Brainerd. *Queens:* throughout, Bisky. *Suffolk:* Miller & Young; Greenport, Tillinghast.
  G. triflorum Michx. f. 3420.
    Kings: Gowanus, Brainerd. *Queens:* Calverley; Flushing, Bisky. *Suffolk:* Miller & Young.
  G. tinctorium L. f. 3423.
    Kings: Prospect Park, Forbell's Landing, Hulst. *Queens:* Flushing, Hempstead, Oyster Bay, Bisky. *Suffolk:* Miller & Young; Greenport, Tillinghast.
  G. concinnum T. & G. f. 3427.
    Suffolk: East Hampton, Mrs. L. D. Pychowska.
  G. asprellum Michx. f. 3428.
    Long Island, Hulst. *Queens:* Calverley; Flushing, Jamaica, Oyster Bay, Bisky.? *Suffolk:* Miller & Young.
  G. decurrens Ives.
    Queens: Glen Cove, Coles.

CAPRIFOLIACEAE.

SAMBUCUS L.
  S. Canadensis L. f. 3432.
    Common throughout the island.
  S. nigra laciniata (Mill.) DC.
    Kings: Prospect Park.

VIBURNUM L.
  V. Opulus L. f. 3435.
    Kings: Prospect Park.
  V. acerifolium L. f. 3437.
    Common throughout the island.
  V. dentatum L. f. 3439.
    Common throughout the island.
  V. cassinoides L. f. 3442.
    Queens: Hempstead, Bisky.
  V. nudum L. f. 3443.
    Kings: New Lots, Hulst. *Queens:* Hempstead, Bisky. *Suffolk:* Miller & Young.
  V. Lentago L. f. 3444.
    Frequent throughout the island.
  V. prunifolium L. f. 3445.
    Kings: Prospect Park; Hulst. *Queens:* Flushing, Newtown, N. Hempstead, Oyster Bay, Bisky.

TRIOSTEUM L.
  T. perfoliatum L. f. 3448.
    Queens: Woodhaven, Hulst; Richmond Hill; Oyster Bay, Bisky; Glen Cove, Coles. *Suffolk:* Miller & Young.
  T. angustifolium L. f. 3449.
    Queens: Oyster Bay, Bisky; Glen Cove, Coles.

LINNAEA L.
  L. borealis L. f. 3450.
    Suffolk: Miller & Young.

SYMPHORICARPOS Juss.
  S. racemosus Mx. f. 3451.
    Kings: Prospect Park; Hulst. *Suffolk:* Miller & Young.
  S. Symphoricarpos (L.) MacM. f. 3454.
    Kings: Hulst; cultivated.

LONICERA L.
  L. caprifolium L. f. 3455.
    Queens: Glen Cove, Coles?
  L. sempervirens L. f. 3461.
    Queens: Richmond Hill, Hulst; Flushing, Hempstead, Newtown, Oyster Bay, Bisky.
  L. Tatarica L. f. 3467.
    Kings: Prospect Park. *Queens:* Hicksville, Hulst.

DIERVILLA Moench.
  D. Diervilla (L.) MacM. f. 3469.
    Kings: Prospect Park. *Queens:* Oyster Bay, Bisky; Glen Cove, Coles.

VALERIANACEAE.
VALERIANA L.
> **V. officinalis** L. f. 3474.
>> *Queens:* Newtown, Bisky; Fresh Pond, Rugers.

VALERIANELLA Poll.
> **V. Locusta** (L.) Bettke. f. 3475.
>> *Queens:* Ronan's Well, Dr. Thurber.

DIPSACACEAE.
DIPSACUS L.
> **D. sylvestris** Huds. f. 3481.
>> *Queens:* Fresh Pond, Hulst, "years ago;" Oyster Bay, Bisky; Glen Cove, Coles.

CUCURBITACEAE.
MICRAMPELIS Raf.
> **M. lobata** (Mx.) Greene. f. 3487.
>> *Kings:* New Lots; Hulst.

SICYOS L.
> **S. angulatus** L. f. 3489.
>> Common throughout the island.

CAMPANULACEAE.
CAMPANULA L.
> **C. rotundifolia** L. f. 3491.
>> *Suffolk:* Miller & Young.
>
> **C. rapunculoides** L. f. 3492.
>> *Queens:* Flushing, Oyster Bay, Bisky.
>
> **C. aparinoides** Pursh. f. 3494.
>> *Kings:* New Lots, Brainerd; Forbell's Landing, Hulst. *Queens:* Calverley; Richmond Hill; Flushing, Hempstead, Jamaica, Oyster Bay, Bisky.
>
> **C. Americana** L. f. 3496.
>> *Queens:* Flushing, Bisky.

LEGOUZIA Durand.
> **L. perfoliata** (L.) Britton. f. 3498.
>> Frequent throughout the island.

LOBELIA L.
> **L. Dortmanna** L. f. 3500.
>> *Suffolk:* Miller & Young.
>
> **L. cardinalis** L. f. 3502.
>> Frequent throughout the island.

L. syphilitica L. f. 3503.
  Kings: New Lots, Brainerd ; Forbell's Landing, Hulst. *Queens:* Flushing, Hempstead, Oyster Bay, Bisky.

L. spicata Lam. f. 3507.
  Frequent throughout the island.

L. inflata L. f. 3509.
  Common throughout the island.

L. Kalmii L. f. 3510.
  Kings · New Lots, Brainerd.

L. Nuttallii Roehm & Schult. f. 3511.
  Queens: Hempstead, Jamaica, Bisky. *Suffolk:* Miller & Young.

L. Canbyi Gray. f. 3512.
  Queens: Cedarhurst, Hulst.

CICHORIACEAE.

CICHORIUM L.
  C. Intybus L. f. 3513.
  Common throughout the county.

ADOPOGON Neck.
  A. Virginicum (L.) Kuntze. f. 3516.
  Kings: Cypress Hills, Hulst. *Queens:* Ridgewood, Glen Cove, Brainerd ; Newtown, Bisky.

  A. Carolinianum (Walt.) Britton. f. 3519.
  Queens: Wood Haven, Hulst ; Hempstead, Newtown, N. Hempstead, Oyster Bay, Bisky. *Suffolk:* Greenport, Tillinghast ; E. Hampton, Mrs. L. D. Pychowska ; Miller & Young.

LEONTODON L.
  L. autumnale L. f. 3522.
  Kings: Cypress Hills, Hulst.

  L. nudicaule (L.) Porter. f. 3523.
  Queens: Cold Spring, Hulst.

TRAGOPOGON L.
  T. pratensis L. f. 3528.
  Queens: Flushing, Bisky.

TARAXACUM Hall.
  T. Taraxacum (L.) Karst. f. 3532.
  Common throughout the island.

SONCHUS L.
  S. arvensis L. f. 3534.
  Queens: Cold Spring, Hulst ; Flushing, Hempstead, Newtown, Bisky.

  S. oleraceus L. f. 3535.
  Kings: Cypress Hills, Hulst. *Queens:* Flushing, Newtown, Bisky. *Suffolk:* Miller & Young.

S. asper (L.) All. f. 3536.
: Queens :* Cold Spring, Hulst ; Newtown, Oyster Bay, Bisky. *Suffolk :* Greenport, Tillinghast.'

LACTUCA L.
L. Scariola L.. f. 3537.
Kings : Brooklyn, Hulst. *Suffolk :* Sag Harbor, Mrs. L. D. Pychowska.
L. Canadensis L.. f. 3539.
Queens : Flushing, Hempstead, N. Hempstead, Oyster Bay, Bisky. *Suffolk :* Greenport, Tillinghast ; Miller & Young.
L. hirsuta Muhl. f. 3540.
Queens : Flushing, Hempstead, Oyster Bay, Bisky. *Suffolk :* Miller & Young.
L. pulchella (Pursh) DC. f. 3542.
Suffolk : Miller & Young.
L. spicata (Lam.) Hitch. f. 3545.
Queens : West Flushing, Rudkin, Newtown, Oyster Bay, Bisky. *Suffolk :* Miller & Young.

HIERACIUM L.
H. aurantiacum L.. f. 3564.
Kings : Cypress Hills, Hulst. *Queens :* W. H. Rudkin.
H. venosum L.. f. 3567.
Frequent throughout the island.
H. Marianum Willd. f. 3568.
Queens : Hempstead, Newtown, Bisky.
H. Canadense Michx. f. 3571.
Common throughout the island.
H. paniculatum L.. f. 3572.
Frequent throughout the island.
H. scabrum Mx. f. 3573.
Frequent throughout the island.
H. Gronovii L. f. 3574.
Frequent throughout the island.

NABALUS Cass.
N. altissimus (L.) Hook. f. 3576.
Kings : Hulst. *Queens :* Richmond Hill ; Flushing, Oyster Bay, Bisky.
N. albus (L.) Hook. f. 3577.
Common throughout the island.
N. serpentarius (Pursh) Hook. f. 3578.
Queens : Flushing, Hempstead, Jamaica, Newtown, Bisky. *Suffolk :* Miller & Young.

N. racemosus (Mx.) DC. f. 3584.
Not infrequent throughout island.

AMBROSIACEAE.
IVA L.
  I. frutescens L. f. 3586.
    Common throughout the island.

AMBROSIA L.
  A. trifida L. f. 3592
    Common throughout the island.
  A. trifida' integrifolia (Muhl) T. & G.
    Frequent.
  A. artemisiaefolia L. f. 3593.
    Common throughout the island.

XANTHIUM L.
  X. spinosum L. 3598.
    Common throughout the island.
  X. strumarium L. f. 3599.
    Frequent throughout the island.
  X. Canadense Mill. f. 3600.
    Frequent throughout the island.

COMPOSITAE.
VERNONIA Schreb.
  V. Noveboracensis (L.) Willd. f. 3601.
    Queens: Hulst; Richmond Hill; throughout, Bisky. *Suffolk*: Miller & Young; Sag Harbor, Mrs. L. D. Pychowska.

EUPATORIUM L.
  E. maculatum L. f. 3614.
    Queens: Plandome, Eddy.
  E. purpureum L. f. 3615.
    Common throughout the island.
  E. leucolepis T. & G. f. 3617.
    Suffolk: Miller & Young.
  E. album L. f. 3618.
    Kings: New Lots, Brainerd. *Suffolk*: Miller & Young.
  E. hyssopifolium L. f. 3619.
    Queens: Hicksville, Hulst; Hempstead, Jamaica, Oyster Bay, Bisky. *Suffolk*: Miller & Young; Sag Harbor, Mrs. L. D. Pychowska.
  E. sessilifolium L. f. 3623.
    Long Island, Hulst. *Queens*: Newtown, Oyster Bay, Bisky; Woodside, Ruger; Glen Cove, Coles. *Suffolk*: Miller & Young.

E. verbenaefolium Mx. f. 3624.
 Kings: Forbell's Landing, Hulst. Queens: Flushing, Jamaica, Hempstead, Bisky. Suffolk: Greenport, Tillinghast; Miller & Young.
E. rotundifolium L. f. 3625.
 Long Island, Hulst. Queens: (Willis' State Flora)? Suffolk: Miller & Young.
E. perfoliatum L. f. 3627.
 Common throughout the island.
E. resinosum Torr. f. 3628.
 Queens: (Willis State Flora).
E. ageratoides L. f. f. 3629.
 Kings: Forbell's Landing, Hulst. Queens: Flushing, N. Hempstead, Oyster Bay, Bisky. Suffolk: Miller & Young.
E. aromaticum L. f. 3630.
 Kings: Coney Island, Brainerd; Cypress Hills, Hulst. Queens: Fresh Pond, Hulst; Hempstead, Jamaica, Bisky; Richmond Hill, Rugers; Hempstead, Allen. Suffolk: Miller & Young.

WILLUGHBAEA Neck.
W. scandens (L.) Kuntze. f. 3632.
 Frequent throughout the island.

LACINARIA Hill.
L. scariosa (L.) Hill. f. 3642.
 Kings: Forbell's Landing, Hulst. Queens: Hempstead, Newtown, Oyster Bay, Bisky. Suffolk: Greenport, Tillinghast; Napeague Beach, Mrs. L. D. Pychowska; Miller & Young.
L. spicata (L.) Kuntze. f. 3643.
 Kings: Forbell's Landing, Hulst. Queens: Jamaica, Brainerd.
L. graminifolia (Walt.) Kuntze. f. 3644.
 Kings: Forbell's Landing, Hulst.

CHRYSOPSIS Nutt.
C. falcata (Pursh) Ell. f. 3653.
 Queens: Cold Spring, D. C. Eaton; Oyster Bay, Bisky. Suffolk: Horton's Point, Southold, Jamesport, Tillinghast, East Hampton, Mrs. L. D. Pychowska; Miller & Young.
C. Mariana (L.) Nutt. f. 3655.
 Kings: Cypress Hills, Hulst; New Lots, Brainerd. Queens: Flushing, Hempstead, Jamaica, Oyster Bay, Bisky. Suffolk: Miller & Young; E. Hampton, Mrs. L. D. Pychowska.

SOLIDAGO L.
S. caesia L. f. 3673.
 Common throughout the island.

S. flexicaulis L. f. 3674.
Kings: Calverley; New Lots, Brainerd. Queens: Oyster Bay, Bisky; Glen Cove, Coles.

S. bicolor L. f. 3676.
Common throughout the island.

S. hispida Muhl. f. 3677.
Queens: Jamaica eastward, Bumstead. Suffolk: Miller & Young.

S. macrophylla Pursh. f. 3680.
Long Island, Hulst. Queens: Sand's Point, Leggett.

S. puberula Nutt. f. 3681.
Queens: Hempstead, Bisky; Bumstead; Rockaway, Leggett. Suffolk: Common, C. H. Peck, 1890, New York State Report; Miller & Young.

S. speciosa Nutt. f. 3685.
Suffolk: East Hampton, Mrs. L. D. Pychowska.

S. speciosa angustata T. & G.
Suffolk: Baiting Hollow, C. H. Peck.

S. sempervirens L. f. 3690.
Common along the beaches throughout the island.

S. odora Ait. f. 3691.
Kings: Prospect Park. Queens: Calverley; Flushing, Hempstead, Jamaica, Oyster Bay, Bisky. Suffolk: Greenport, Tillinghast; East Hampton, Mrs. L. D. Pychowska; Miller & Young.

S. rugosa Mill. f. 3693.
Queens: Calverley; Flushing, Hempstead, Jamaica, N. Hempstead, Bisky.

S. patula Muhl. f. 3695.
Queens: Astoria, Bumstead.

S. ulmifolia Muhl. f. 3696.
Queens: Rockaway, Leggett. Suffolk: Miller & Young.

S. Elliottii T. & G. f. 3698.
Kings: Forbelli Landing, Hulst. Suffolk: Miller & Young.

S. neglecta T. & G. f. 3699.
Queens: Hempstead, Newtown, N. Hempstead, Bisky. Suffolk: Miller & Young.

S. arguta Ait. f. 3702.
Kings: Calverley. Queens: Calverley; Hempstead, Newton, Bisky; Rockaway Beach. Suffolk: Greenport, Tillinghast; Miller & Young; East Hampton, Mrs. L. D. Pychowska.

S. serotina Ait. f. 3704.
Queens: Calverley; Oyster Bay, Bisky.

S. Canadensis L. f. 3708.
Common throughout the island.
S. Canadensis scabriuscula Porter.
Suffolk: Miller & Young.
S. nemoralis Ait. f. 3709.
Common throughout the island.
S. rigida L. f. 3713.
Suffolk: East Hampton, Mrs. L. D. Pychowska.

EUTHAMIA Nutt.
E. graminifolia (L.) Nutt. f. 3718.
Common throughout the island.
E. Caroliniana (L.) Greene. f. 3720.
Common throughout the island.

SERIOCARPUS Nees.
S. linifolius (L.) B.S.P. f. 3732.
Kings: Forbell's Landing, Hulst. *Queens:* Hicksville, Hulst ; Hempstead, Jamaica, Bisky. *Suffolk:* Miller & Young.
S. asteroides (L.) B.S.P. f. 3734.
Kings: Forbell's Landing, Hulst. *Queens:* Jamaica, Brainerd ; Throughout, Bisky. *Suffolk:* East Hampton, Mrs. L. D. Pychowska ; Miller & Young.

ASTER L.
A. divaricatus L. f. 3737.
Common throughout the island.
A. macrophyllus L. f. 3743.
Kings: Forbell's Landing, Hulst. *Queens:* Flushing, Newtown, Oyster Bay, Bisky.
A. cordifolius L. f. 3752.
Common throughout the island.
A. sagittifolius Willd. f. 3756.
Long Island, Hulst. *Queens:* Flushing, Bisky.
A. undulatus L. f. 3757.
Common throughout the island.
A. patens Ait. f. 3758.
Common throughout the island.
A. phlogifolius Muhl. f. 3759.
Queens: Newtown, Bisky ; Mrs. E. G. Britton ; E. N. Day.
A. Novae-Angliae L. f. 3760.
Kings: New Lots, Brainerd ; Forbell's Landing, Hulst. *Queens:* Calverley ; Flushing, Hempstead, N. Hempstead, Oyster Bay, Bisky. *Suffolk:* East Hampton, Mrs. L. D. Pychowska ; Miller & Young.

A. puniceus L.   f. 3764.
   Kings: New Lots, Brainerd.   *Queens:* Throughout, Bisky.
   Suffolk: Miller & Young.

A. prenanthoides Muhl.   f. 3767.
   Long Island, Hulst.

A. laevis L.   f. 3768.
   Queens: Maspeth, Hulst; Flushing, Newtown, Oyster Bay, Bisky; Glen Cove, Coles.   *Suffolk:* Greenport, Tillinghast; Miller & Young.

A. Novi-Belgii L.   f. 3773.
   Kings: Forbell's Landing, Hulst.   *Queens:* Richmond Hill; Flushing, Hempstead, Jamaica, Bisky.   *Suffolk:* East Hampton, Mrs. L. D. Pychowska; Miller & Young.

A. Novi-Belgii elodes (T. & G.) Gray.
   Queens: (Torrey's State Flora.)

A. longifolius Lam.   f. 3774.
   Suffolk: Miller & Young.

A. concolor L.   f. 3777.
   Queens: Woodhaven, Hulst; Rudkin; Ridgewood, Aqueduct, Torrey Club Herb., Ruger.   *Suffolk:* Miller & Young.

A. spectabilis Ait.   f. 3780.
   Kings: New Lots, Brainerd.   *Queens:* Woodhaven, Hulst; Hempstead, Jamaica, Bisky.

A. Radula Ait.   f. 3783.
   Queens: Hempstead, Bisky.

A. nemoralis Ait.   f. 3788.
   Kings: Bay Ridge, Brainerd.

A. acuminàtus Mx.   f. 3789.
   Common throughout the island.

A. dumosus L.   f. 3791.
   Kings: New Lots, Brainerd.   *Queens:* Hempstead, Jamaica, Bisky.

A. paniculatus Lam.   f. 3793.
   Kings: New Lots, Brainerd.   *Queens:* Calverley; Hempstead, Bisky; Eaton, Bumstead.   *Suffolk.*

A. Tradescanti L.   f. 3795.
   Common throughout.

A. ericoides L.   f. 3797.
   Queens: Throughout, Bisky.   *Suffolk:* Greenport, Tillinghast; E. Hampton, Mrs. L. D. Pychowska; Miller & Young.

A. lateriflorus (L.) Britton.   f. 3799.
   Common throughout the island.

A. vimineus Lam. f. 3801.
Kings: Forbell's Landing, Hulst. *Queens.* *Suffolk:* C. H. Peck. New York State Report, 1890 ; Miller & Young.

A. multiflorus Ait. f. 3802.
Kings: Gowanns, Brainerd. *Queens:* Hempstead, Newtown, Oyster Bay, Bisky. *Suffolk:* Miller & Young.

A. tenuifolius L. f. 3804.
Queens: Calverley; Hempstead, Newtown, Bisky. *Suffolk:* Greenport, Tillinghast ; Miller & Young.

A. subulatus Mx. f. 3806.
Kings: Forbell's Landing, Hulst. *Queens:* Flushing, Hempstead, Jamaica, Newtown, Bisky. *Suffolk:* Miller & Young.

ERIGERON L.
E. pulchellus Mx. f. 3819.
Common throughout the island.

E. Philadelphicus L. f. 3820.
Queens: Oyster Bay, Bisky; Glen Cove, Coles. *Suffolk:* Miller & Young.

E. annuus (L.) Pers. f. 3823.
Queens: Throughout, Bisky. *Suffolk:* Greenport, Tillinghast; Miller & Young.

E. ramosus (Walt.) B.S.P. f. 3824.
Common throughout the island.

LEPTILON Raf.
L. Canadensis (L.) Britton. f. 3827.
Common throughout the island.

DOELLINGERIA Nees.
D. umbellatus (Mill.) Nees. f. 3829.
Kings: New Lots, Brainerd ; Forbell's Landing, Hulst. *Queens:* Flushing, Hempstead, Oyster Bay, Bisky. *Suffolk:* Miller & Young ; East Hampton, Mrs. L. D. Pychowska.

D. infirma (Mx.) Greene. f. 3831.
Queens: Astoria, Leggett.

IONACTIS Greene.
I. linariifolius (L.) Greene. f. 3832.
Common throughout the island.

BACCHARIS L.
B. halimifolia L. f. 3834.
Common throughout the island.

PLUCHEA Cass.
P. camphorata (L.) DC. f. 3840.
Common along the coasts of the island.

ANTENNARIA Gaertn.
  A. plantaginifolia (L.) Richards. f. 3848.
    Common throughout the island.

ANAPHALIS DC.
  A. margaritacea (L.) B. & H. f. 3850.
    Common throughout the island.

GNAPHALIUM L.
  G. obtusifolium L. f. 3851.
    Common throughout the island.
  G. uliginosum L. f. 3855.
    Frequent throughout the island.
  G. purpureum L. f. 3859.
    Queens: Hempstead, Bisky. *Suffolk:* Greenport, Tillinghast; Miller & Young.

INULA L.
  I. Helenium L. f. 3861.
    Queens: Flushing, Newtown, Oyster Bay, Bisky. *Suffolk:* Miller & Young.

SILPHIUM L.
  S. perfoliatum L. f. 3865.
    Queens: Flushing, Bisky.

HELIOPSIS Pers.
  H. helianthoides (L.) B.S.P. f. 3878.
    Kings: Forbell's Landing, Hulst.

ECLIPTA L.
  E. alba (L.) Hassk. f. 3880.
    Long Island, Hulst.

RUDBECKIA L.
  R. hirta L. f. 3885.
    Common throughout the island.
  R. fulgida Ait. f. 3887.
    Kings: Cypress Hills, Hulst.
  R. laciniata L. f. 3890.
    Kings: Ridgewood, Brainerd, Prospect Park. *Queens:* Flushing, Bisky.

HELIANTHUS L.
  H. angustifolius L. f. 3898.
    Kings: New Lots, Brainerd; Forbell's Landing, Hulst. *Queens:* Hempstead, Newton, Oyster Bay, Bisky. *Suffolk:* Miller & Young.
  H. annuus L. f. 3900.
    Common throughout the island.

H. giganteus L. f. 3907.
Common throughout the island.

H. divaricatus L. f. 3910.
Common throughout the island.

H. decápetalus L. f. 3913.
Kings: New Lots, Brainerd. *Queens:* Newtown, Bisky.

H. strumosus L. f. 3915.
Kings: Forbell's Landing, Hulst. *Queens:* Calverley; Newtown, Bisky. *Suffolk:* Greenport, Tillinghast; Miller & Young.

H. tuberosus L. f. 3919.
Kings: Forbell's Landing, Hulst. *Queens:* Flushing, Hempstead, Jamaica, Bisky. *Suffolk:* Miller & Young.

COREOPSIS L.
C. rosea Nutt. f. 3925.
Suffolk: Miller & Young.

BIDENS L.
B. laevis (L.) B.S.P. f. 3938.
Kings: Forbell's Landing, Hulst. *Queens:* Flushing, Hempstead, Jamaica, N. Hempstead, Bisky. *Suffolk:* Miller & Young.

B. cernua L. f. 3939.
Kings: Prospect Park; Forbell's Landing, Hulst. *Queens:* Bisky.

B. connata Muhl. f. 3940.
Common throughout the island.

B. discoidea (T. & G.) Britton. f. 3943.
Queens: Jamaica, Bisky.

B. frondosa L. f. 3944.
Common throughout the island.

B. bipinnata L. f. 3945.
Frequent throughout the island.

B. trichosperma (Mx.) Britton. f. 3947.
Kings: Forbell's Landing, Cedarhurst, Hulst. *Queens:* Hempstead, Bisky.

B. trichosperma tenuiloba (Gray) Britton.
Suffolk: Eastport, Patchogue, C. H. Peck, New York State Reports, 1889.

GALINSOGA R. & P.
G. parviflora Cav. f. 3954.
Common throughout western part of island.

HELENIUM L.
   H. autumnale L.. f. 3972.
   Common throughout the island.

ACHILLEA L..
   A. Millefolium L. f. 3983.
   Common throughout the island.

ANTHEMIS L.
   A. Cotula L. f. 3984.
   Common throughout the island.
   A. arvensis L. f. 3985.
   Kings: Calverley; Forbell's Landing, Hulst. *Queens:* Woodhaven, Hulst; Flushing, Bisky.

CHRYSANTHEMUM L.
   C. Leucanthemum L. f. 3988.
   Common throughout the island.
   C. Parthenium (L.) Pers. f. 3990.
   Queens: Cold Spring, Hulst; Jamaica, Newtown, Oyster Bay, Bisky. *Suffolk:* Greenport, Tillinghast; Miller & Young.

MATRICARIA L.
   M. inodora L. f. 3992.
   Queens: Newtown, Bisky. *Suffolk:* Miller & Young.
   M. Chamomilla L. f. 3994.
   Kings: Calverley.

TANACETUM L..
   T. vulgare L. f. 3996.
   Common throughout the island.

ARTEMISIA L.
   A. caudata Michx. f. 3998.
   Queens: Cold Spring, Hulst; Glen Cove, Coles. *Suffolk:* Miller & Young.
   A. biennis Willd. f. 4008.
   Kings: Ballast, Brooklyn, Hulst.
   A. Stellariana Besser. f. 4009.
   Kings: Forbell's Landing, Hulst.
   A. vulgaris L. f. 4010.
   Kings: Brooklyn, Hulst. *Queens:* Glen Cove, Coles.

TUSSILAGO L.
   T. Farfara L. f. 4019.
   Queens: (Roslyn?), Bisky. *Suffolk:* Miller & Young.

ERECHTITES Raf.
   E. hieracifolia (L.) Raf. f. 4028.
   Common throughout the island.

MESADENIA Raf.
    M. atriplicifolia (L.) Raf. f. 4030.
        Suffolk: Miller & Young.

SENECIO L.
    S. Balsamitae Muhl. f. 4043.
        Kings: Prospect Park. *Queens:* Jamaica, Bisky.
    S. aureus L. f. 4047.
        Common throughout the island.
    S. vulgaris L. f. 4053.
        Kings: New Lots, Hulst; Gravesend. *Queens:* Cold Spring, Hulst; Flushing, Newtown, Bisky.

ARCTIUM L.
    A. Lappa L. f. 4056.
        Common throughout the island.

CARDUUS L.
    C. lanceolatus L. f. 4058.
        Common throughout the island.
    C. discolor (Muhl.) Nutt. f. 4060.
        Frequent throughout.
    C. odoratus (Muhl.) Porter. f. 4067.
        Queens: Flushing, Bisky. *Suffolk:* Greenport, Tillinghast; Miller & Young.
    C. spinosissimus Walt. f. 4069.
        Kings: Forbell's Landing, Hulst. *Queens:* Hempstead, Oyster Bay, Bisky. *Suffolk:* Greenport, Tillinghast; Miller & Young; E. Hampton, Mrs. L. D. Pychowska.
    C. muticus (Mx.) Pursh. f. 4070.
        Kings: Forbell's Landing, Hulst. *Queens:* Flushing, Bisky. *Suffolk:* Miller & Young.
    C. arvensis (L.) Robs. f. 4071.
        Common throughout the island.
    C. nutans L. f. 4072.
        Kings: Brooklyn, Hulst.

ONOPORDON L.
    O. Acanthium L. f. 4075.
        Kings: Brooklyn, Hulst. *Queens:* Flushing, Bisky. *Suffolk:* Miller & Young.

CENTAUREA L.
    C. Cyanus L. f. 4076.
        Kings: Brooklyn, Hulst.

INDEX OF GENERA.

Abies, 60
Abutilon, 115
Acalypha, 112
Acaulon, 49
Acer, 114
Acerates, 129
Achillea, 153
Achnanthes, 22
Achroanthes, 81
Acnida, 90
Acorus, 74
Actaea, 94
Actinocyclus, 20
Actinoptychus, 20
Adiantum, 58
Adicea, 86
Adopogon, 143
Aecidium, 36
Aesculus, 114
Aethusa, 121
Agastache, 132
Agrimonia, 103
Agropyron, 68
Agrostemma, 91
Agrostis, 66
Ahnfeldtia, 12
Ailanthus, 111
Aira, 66
Alaria, 11
Albugo, 27
Alectoria, 46
Aletris, 78
Alisma, 62
Allium, 77
Alnus, 84
Alopecurus, 65
Alsine, 92
Altenaria, 33
Althaea, 115
Amanita, 42
Amaranthus, 90
Amblystegium, 56
Ambrosia, 145
Amelanchier, 104
Ammodenia, 92
Ammophila, 66
Amorpha, 107
Amphiprora, 24
Amphisphaeria, 30
Amphora, 25
Amygdalus, 105
Anagallis, 126

Anaphalus, 151
Andropogon, 63
Anemone, 95
Angelica, 120
Anomodon, 54
Antennaria, 151
Anthemis, 153
Anthoceras, 49
Anthoxanthum, 64
Anychia, 93
Apios, 109
Apocynum, 128
Aposphaeria, 33
Aptogonum, 3
Aquilegia, 94
Arabis, 99
Aralia, 120
Arctium, 154
Arctostaphylos, 125
Arcyria, 1
Arenaria, 92
Arethusa, 80
Argemone, 97
Arisaema, 74
Aristida, 64
Aristolochia, 86
Armillaria, 42
Aronia, 104
Artemisia, 153
Arthocladia, 9
Arthrodesmus, 2
Asarum, 86
Asclepias, 128
Ascophyllum, 11
Ascyrum, 116
Asparagus, 78
Aspergillus, 29
Asplenium, 58
Aster, 148
Asterionella, 22
Atriplex, 89
Aulacomnion, 53
Auliscus, 20
Azalea, 124

Baccharis, 150
Bacillus, 27
Bacterium, 26
Baeomyces, 43
Bangia, 11
Baptisia, 106
Barbarea, 98

Barbula, 51
Bartonia, 128
Batrachium, 96
Batrachospermum, 12
Bartramia, 53
Beggiatoa, 27
Benzoin, 97
Berberis, 96
Betula, 83
Biatora, 44
Biddulphia, 20
Bidens, 152
Blepharostoma, 48
Boehmeria, 86
Boletus, 39
Bostrychia, 14
Botrychium, 57
Brachyelytrum, 65
Brachythecium, 55
Brasenia, 93
Brassica, 98
Bromus, 68
Broussonetia, 85
Bryopsis, 8
Bryum, 52
Buellia, 44
Bulbochaete, 7
Bulbocoleon, 6
Bursa, 99
Butnera, 96

Cakile, 97
Calamagrostis, 66
Calla, 74
Collinsonia, 134
Callithamnion, 15
Callitriche, 112
Calothrix, 18
Caltha, 94
Camelina, 99
Campanula, 142
Cannabis, 86
Cantharellus, 40
Capriola, 66
Cardamine, 98
Carduus, 154
Carex, 71
Carpinus, 83
Carum, 122
Cassia, 105
Castagnea, 10
Castalia, 93

Castanea, 84
Castilleja, 137
Catalpa, 138
Catharinea, 53
Ceanothus, 115
Celastrus, 114
Celtis, 85
Cenchrus, 64
Centaurea, 154
Cephalanthus, 139
Cephalozia, 48
Ceratium, 17
Cerastium, 92
Ceramium, 16
Ceratiomyxa, 1
Ceratodon, 50
Ceratophyllum, 93
Cercis, 105
Cercospora, 32
Cetraria, 45
Chaerophyllum, 121
Chaetomium, 30
Chaetomorpha, 7
Chaetophora, 6
Chaetosphaeria, 30
Chamaecyparis, 60
Chamaedaphne, 124
Chamaelirium, 76
Chamaenirion, 119
Champia, 13
Chelone, 136
Chelidonium, 97
Chenopodium, 88
Chiloscyphus, 48
Chimaphila, 123
Chiogenes, 125
Chionanthus, 127
Chlamydomonas, 4
Chondria, 13
Chondrus, 12
Chorda, 11
Chordaria, 10
Chrosperma, 76
Chrysanthemum, 153
Chrysopsis, 146
Chrysosplenium, 101
Cichorium, 143
Cicuta, 122
Cimicifuga, 94
Cinna, 65
Circaea, 120
Circinella, 28
Cladium, 71
Cladonia, 43
Cladophora, 7
Cladosporium, 32
Cladostephus, 9
Cladothrix, 27
Cladrastris, 106
Clathrocystis, 17
Clavaria, 37

Claytonia, 90
Climacium, 55
Clitocybe, 41
Clitoria, 108
Clematis, 95
Cleome, 99
Clethra, 123
Closterium, 2
Cocconeis, 22
Coelastrum, 5
Coelosphaerium, 17
Coleosporium, 34
Collenia, 46
Collybia, 41
Comandra, 86
Comatricha, 1
Commelina, 75
Comptonia, 82
Conferva, 6
Conium, 121
Conocephalum, 47
Conotrema, 46
Convallaria, 78
Convolvulus, 129
Coprinus, 41
Corallina, 17
Corallorhiza, 81
Cordyceps, 30
Corema, 113
Coreopsis, 152
Cornus, 122
Coronilla, 107
Coronopus, 97
Cortinarius, 40
Corylus, 83
Coscinodiscus, 20
Cosmarium, 2
Cracca, 107
Crataegus, 104
Crotalaria, 106
Craterellus, 37
Crucibulum, 42
Cunila, 133
Cuscuta, 130
Cyathus, 42
Cyclomyces, 39
Cyclotella, 19
Cylindrothecium, 55
Cymatopleura, 26
Cymbella, 25
Cynanchum, 129
Cynoglossum, 130
Cyperus, 69
Cypripedium, 79
Cystoclonium, 12
Cytisus, 106

DACTYLIS, 67
Daedalia, 38
Daldinea, 31
Danthonia, 66

Darluca, 33
Dasya, 14
Dasystoma, 137
Datura, 135
Daucus, 120
Decoden, 118
Delesseria, 13
Derniga, 122
Delphinium, 94
Dermocybe, 40
Deschampsia, 66
Desmarestia, 9
Desmidium, 3
Dentaria, 99
Dianthus, 91
Diatoma, 21
Diatrype, 31
Dichelyma, 54
Dicksonia, 57
Dicladia, 26
Dicranella, 50
Dicranum, 50
Dictyosiphon, 10
Dictydium, 1
Dictyosphaerium, 5
Diervilla, 141
Digitalis, 137
Diodia, 140
Dioscorea, 79
Diospyros, 127
Diphyscium, 54
Diplotaxis, 98
Dipsacus, 142
Dirca, 118
Distichlis, 67
Ditrichum, 50
Docidium, 2
Dodecatheon, 126
Doellingeri, 150
Dondia, 89
Dothidea, 30
Draba, 99
Draparnaldia, 6
Drosera, 100
Drummondia, 52
Dryopteris, 58
Duchesnia, 102
Dulichium, 69

EATONIA, 6
Echinobotrys, 32
Echium, 131
Eclipta, 151
Ectocarpus, 8
Elachista, 10
Eleocharis, 69
Eleusine, 67
Elymus, 68
Encyonema, 25
Endocarpon, 46
Entoloma, 41

Epichloe, 30
Epigaea, 125
Epilobium, 119
Eragrostis, 67
Erechtites, 153
Erigeron, 150
Eriocaulon, 75
Eriophorum, 71
Erodium, 110
Erysibe, 29
Erythraea, 127
Erythronium, 77
Euastrum, 3
Eunotia, 22
Eunotogramma, 21
Euonymus, 114
Eupodiscus, 20
Eupatorium, 145
Euphorbia, 112
Euthamia, 148
Euthora, 13
Eutypa, 31
Eutypella, 31
Exidia, 36
Exobasidium, 36

FAGOPYRUM, 87
Fagus, 84
Falcata, 109
Favolus, 38
Festuca, 68
Ficaria, 96
Fimbristylis, 70
Fissidens, 50
Fistulina, 39
Fontinalis, 54
Fragaria, 102
Fragilaria, 22
Fraxinus, 127
Frullania, 48
Fucus, 11
Fuirena, 71
Fuligo, 2
Funaria, 52
Fusarium, 33
Fusicladium, 32

GALACTIA, 110
Galeopsis, 132
Galinsoga, 152
Galium, 140
Gaultheria, 125
Gaylussacia, 125
Geaster, 42
Gelidium, 12
Gemmingia, 79
Genista, 106
Gentiana, 127
Georgia, 52
Geoglossum, 28
Geranium, 110

Gerardia, 137
Geum, 103
Gigartina, 12
Glaucium, 97
Glecoma, 132
Gleditsia, 105
Gnaphalium, 15
Gomphonema, 25
Gonium, 4
Gracilaria, 13
Grammatophora, 21
Graphiola, 34
Graphis, 44
Gratiola, 136
Griffithsia, 15
Grimmia, 51
Grinnellia, 13
Gymnocladus, 106
Gymnodinium, 17
Gymnogongrus, 12
Gymnosporangium, 34
Gyrostachys, 80
Gyrotheca, 79

HABENARIA, 80
Halurus, 15
Hamamelis, 101
Hedeoma, 133
Hedwigia, 51
Helenium, 153
Heleochloa, 65
Heliantheum, 116
Helianthus, 151
Heliopsis, 151
Helleborus, 94
Helotium, 28
Hemerocallis, 77
Hemiarcyria, 1
Hemicarpha, 71
Hendersonia, 31
Hepatica, 95
Heracleum, 121
Hesperis, 99
Heuchera, 101
Hibiscus, 115
Hicorea, 82
Hieracium, 144
Hildenbrandtia, 17
Holacanthum, 2
Holcus, 66
Homalocenchrus, 64
Hottonia, 126
Houstonia, 139
Hudsonia, 117
Humulus, 86
Hyalodiscus, 19
Hyalotheca, 3
Hydnum, 37
Hydrocotyle, 122
Hydrodictyon, 5
Hygrophorus, 40

Hylocomium, 56
Hymenochaete, 36
Hymenostylium, 49
Hyoscyamus, 135
Hypericum, 116
Hypholoma, 41
Hypnea, 13
Hypnum, 56
Hypocrea, 30
Hypomyces, 30
Hypopytis, 123
Hypoxis, 79
Hypoxylon, 31
Hysterium, 29

ILEX, 113
Ilicioides, 113
Ilysanthes, 136
Impatiens, 114
Inula, 151
Ionactis, 150
Ipomoea, 129
Iris, 79
Irpex, 37
Isactis, 19
Isanthes, 131
Isaria, 33
Isnardia, 119
Isthmia, 21
Iva, 145
Ixophorus, 64

JUGLANS, 81
Juncoides, 76
Juncus, 75
Juniperus, 60

KALMIA, 124
Kantia, 48
Kneiffia, 119
Koellia, 133
Koniga, 99
Kosteletzkya, 115

LACINARIA, 146
Lactarius, 40
Lactuca, 144
Lappula, 130
Laminaria, 11
Lamium, 132
Larix, 60
Lathyrus, 108
Leathesia, 10
Lecanora, 45
Lechea, 117
Lecidea, 44
Legouzia, 142
Lemna, 74
Lenzites, 38
Leontodon, 143
Leonurus, 132

Leotia, 28
Lepidium, 97
Lepedozia, 48
Leptamnium, 138
Leptandra, 137
Leptilon, 150
Leptobryum, 52
Leptogium, 46
Leptorchis, 81
Leptostroma, 34
Lespedeza, 108
Leucobryum, 50
Leucodon, 54
Leucothoë, 124
Ligustrum, 127
Lilium, 77
Limnanthemum, 128
Limodorum, 81
Limonium, 126
Linaria, 135
Linnaea, 141
Linum, 111
Liquidamber, 101
Liriodendron, 94
Lithospermum, 130
Lobelia, 142
Lomentaria, 13
Lonicera, 141
Lophoclea, 48
Lophodermium, 29
Ludwigia, 119
Lupinus, 106
Lycium, 135
Lycogala, 1
Lycoperdon, 42
Lycopersicon, 135
Lycopodium, 59
Lycopsis, 131
Lycopus, 133
Lychnis, 91
Lyngbya, 18
Lysimachia, 126
Lythrum, 118

MACROSPORIUM, 33
Magnolia, 94
Malus, 104
Malva, 115
Marasmius, 39
Marchantia, 47
Marrubium, 132
Martynia, 139
Massaria, 30
Matricaria, 153
Medeola, 78
Medicago, 106
Meibomia, 108
Melampyrum, 137
Melanconium, 34
Melanthium, 76
Melilotus, 106

Melissa, 133
Melobesia, 16
Melosira, 19
Menispermum, 96
Mentha, 134
Menyanthes, 128
Meridion, 21
Merismopedia, 17
Mesadenia, 154
Mesogloia, 10
Micrampelis, 142
Micrasterias, 3
Micrococcus, 26
Microcoleus, 18
Microsphaera, 29
Mimulus, 136
Mitchella, 140
Mitella, 101
Mnium, 52
Mohringia, 92
Molluga, 90
Monarda, 133
Monilia, 32
Monostroma, 5
Monotropa, 123
Morus, 85
Morchella, 28
Mucor, 27
Mucronoporus, 37
Muhlenbergia, 65
Muscari, 77
Mutinus, 43
Mycena, 41
Myosotis, 130
Myriactis, 10
Myrica, 82
Myrionema, 10
Myriophyllum, 120
Myriotrichia, 10

NABALUS, 144
Naias, 62
Naucoria, 41
Naumbergia, 126
Navicula, 23
Nectria, 30
Nemalion, 12
Nelumbo, 93
Nepeta, 132
Nitzchia, 25
Nodularia, 18
Nostoc, 17
Notothylas, 49
Nummularia, 31
Nymphaea, 93
Nyssa, 123

ODONTOSCHISMA, 48
Oldenlandria, 139
Omphalia, 41
Onagra, 119

Onoclea, 57
Onopordon, 154
Onosmodium, 131
Opegrapha, 44
Ophioglossum, 57
Opulaster, 102
Opuntia, 118
Orchis, 80
Ornithogalum, 77
Orontium, 74
Orthotrichum, 51
Oscillaria, 18
Osmunda, 57
Ostrya, 83
Oxalis, 110
Oxycoccus, 125
Oxygraphis, 96
Oxypolis, 121

PALLAVICINIA, 47
Panax, 120
Pandorina, 4
Panicularia, 67
Panicum, 63
Panus, 39
Paramium, 81
Parmelia, 45
Parnassia, 101
Parsonia, 118
Parthenocissus, 115
Paspalum, 63
Pastinaca, 121
Paulownia, 136
Pediastrum, 5
Pedicularis, 137
Pellia, 47
Peltandra, 74
Peltigera, 46
Penicillium, 29
Penium, 2
Penthorum, 100
Pentstemon, 136
Peridinium, 17
Pertussaria, 46
Petunia, 135
Peysonnelia, 16
Plagiothecium, 55
Phalaris, 64
Phallus, 43
Phaseolus, 110
Phegopteris, 58
Philonotus, 53
Philotria, 62
Phleum, 65
Phlox, 130
Pholiota, 41
Phoma, 33
Phragmidium, 35
Phryma, 139
Phyllachora, 30
Phyllitis, 9

Phyllophora, 12
Phyllosticta, 33
Physalis, 134
Physalodes, 134
Physostegia, 132
Physcia, 44
Physcomitrium, 52
Phytolacca, 90
Picea, 60
Pieris, 124
Pimpinella, 121
Pinus, 59
Placodium, 44
Plagiochila, 47
Plagiogramma, 22
Plantago, 139
Platanus, 101
Platygyrium, 54
Pleonectria, 50
Pleuridium, 49
Pleurococcus, 5
Pleurostigma, 24
Pleurotaenium, 2
Pleurotus, 41
Plowrightia, 30
Pluchea, 30
Plumaria, 16
Poa, 67
Pogonatum, 53
Pogonia, 80
Polanisia, 100
Polyedrium, 5
Polygala, 111
Polygonatum, 78
Polygonella, 88
Polygonum, 87
Polyides, 16
Polypodium, 59
Polyporus, 38
Polysiphonia, 14
Polytrichum, 53
Pontederia, 75
Populus, 82
Porella, 48
Portulaca, 90
Porphyra, 11
Potamogeton, 61
Potentilla, 103
Pottia, 51
Proserpinaca, 120
Protomyces, 28
Prunella, 132
Prunus, 105
Psalliota, 41
Pseudomonas, 27
Psilocarya, 70
Ptelea, 111
Pteris, 59
Ptilimnium, 122
Ptilota, 16
Puccinia, 35

Punctaria, 9
Pylaisia, 55
Pyrola, 123
Pyxilla, 26
Pyxine, 44

QUERCUS, 84

RACOMITRIUM, 51
Radula, 48
Ramalina, 45
Ranunculus, 95
Raphanus, 98
Raphidium, 5
Reseda, 100
Reticularia, 1
Rhabdonema, 21
Rhabdonia, 13
Rhamnus, 114
Rhaphidostegium, 55
Rhexia, 119
Rhizoclonium, 7
Rhizopus, 28
Rhizosolenia, 20
Rhodobryum, 52
Rhododendron, 124
Rhodomela, 14
Rhodora, 124
Rhodymenia, 13
Rhoicosphenia, 25
Rhus, 113
Rhytisma, 29
Ribes, 101
Riccardia, 47
Riccia, 47
Ricinus, 112
Rivularia, 19
Robinia, 107
Roestelia, 35
Roripa, 98
Rosa, 103
Rotala, 118
Rubus, 102
Rudbeckia, 151
Rumex, 87
Ruppia, 62
Russula, 40
Ruta, 111
Rynchospora, 71

SABBATIA, 127
Sacidium, 34
Saccharomyces, 28
Saccogyna, 48
Sagina, 92
Sagittaria, 62
Salicornia, 89
Salix, 83
Salsola, 89
Sambucus, 140

Samolus, 126
Sanguinaria, 97
Sanguisorba, 103
Sanicula, 121
Saponaria, 91
Sarcina, 26
Sargassum, 11
Sarracenia, 100
Sarothra, 116
Sassafras, 96
Saururus, 81
Savastana, 64
Saxifraga, 100
Scapania, 48
Scenedesmus, 5
Schizophyllum, 39
Scinaia, 12
Scirpus, 70
Scleranthus, 93
Scleria, 71
Scleroderma, 42
Scoliopleura, 24
Scropularia, 136
Scutellaria, 131
Sedum, 100
Selaginella, 59
Senecio, 154
Septonema, 32
Septoria, 34
Seriocarpus, 148
Sesuvium, 90
Sicyos, 142
Silene, 91
Silphium, 151
Sinapis, 97
Sisymbrium, 97
Sisyrinchium, 79
Sium, 122
Smilax, 78
Solanum, 135
Solidago, 146
Sonchus, 143
Sorbus, 104
Sparganium, 61
Spartina, 66
Spathyema, 74
Spergula, 93
Spermothamnion, 14
Sphacelaria, 9
Sphaerella, 4
Sphaeropsis, 33
Sphaerotheca, 29
Sphaerozosma, 3
Sphaerozyga, 18
Sphagnum, 49
Spiraea, 102
Spirodela, 74
Spirogyra, 4
Spirotaenia, 2
Spirulina, 18
Sporidesmium, 32

Sporobolus, 65
Sporocybe, 33
Sporotrichum, 32
Spyridia, 16
Stachyobotrys, 32
Stachys, 133
Staphylea, 114
Staurastrum, 3
Stauroneis, 24
Steironema, 126
Stemonitis, 1
Stenophyllus, 70
Stereodon, 56
Stereum, 36
Stichococcus, 5
Sticta, 47
Stilophora, 10
Stipa, 65
Streptococcus, 26
Striaria, 9
Striatella, 21
Strobilomyces, 39
Strophostyles, 110
Stylosanthes, 107
Surirella, 26
Symphoricarpos, 141
Symphytum, 131
Synchytrium, 27
Syndesmon, 95
Synedra, 22
Syntherisma, 63
Syringa, 127

TABELLARIA, 21
   Tanacetum, 153
Taraxacum, 143
Taxus, 60
Tecoma, 138
Terpsinoe, 21
Tetraspora, 4
Teucrium, 131
Thalesia, 138
Thalictrum, 96
Thaspium, 121
Thelephora, 37

Thelia, 56
Theloschistes, 44
Thlaspi, 97
Tilia, 115
Tissa, 93
Thuidium, 54
Thuja, 60
Toxylon, 85
Tradescantia, 75
Trametes, 38
Trapopogon, 143
Tremella, 36
Trentepohlia, 6
Triadenum, 116
Triceratium, 20
Trichia, 1
Trichocolea, 48
Tricholoma, 42
Trichostema, 131
Trientalis, 126
Trifolium, 107
Triglochin, 62
Trillium, 78
Triosteum, 141
Tripsacum, 63
Trollius, 94
Tsuga, 60
Tubercularia, 33
Tubulina, 1
Tunica, 91
Tussilago, 153
Typha, 61

ULMUS, 85
   Ulota, 51
Ulothrix, 6
Ulva, 6
Uncinula, 29
Unifolium, 78
Uredo, 36
Uromyces, 35
Urtica, 86
Urticastrum, 86
Usnea, 46
Ustilago, 34

Ustulina, 31
Utricularia, 138
Uvularia, 77

VACCARIA, 91
   Vaccinium, 125
Vagnera, 78
Valeriana, 142
Valerianella, 142
Vallisneria, 62
Valsa, 31
Vaucheria, 8
Veratrum, 77
Verbascum, 135
Verbena, 131
Vernonia, 145
Veronica, 136
Viburnum, 141
Vicia, 108
Vinca, 128
Viola, 117
Vitis, 115
Volvox, 4

WASHINGTONIA, 121
   Weisia, 49
Willughbaea, 146
Woodsia, 57
Woodwardia, 58

XANTHIDIUM, 3
   Xanthium, 145
Xanthorrhiza, 94
Xanthoxylum, 111
Xolisma, 124
Xylaria, 32
Xyris, 75

ZANNICHELLIA, 62
   Zizania, 64
Zizia, 122
Zostera, 62
Zygnema, 4
Zygogonium, 4

www.ingramcontent.com/pod-product-compliance
Lightning Source LLC
Chambersburg PA
CBHW030402230426
43664CB00007BB/710